HISTORY'S
GREATEST
DISCOVERIES

HISTORY'S
GREATEST
DISCOVERIES

AND THE PEOPLE WHO MADE THEM

JOEL LEVY

METRO BOOKS
New York

METRO BOOKS
New York

An Imprint of Sterling Publishing
1166 Avenue of the Americas
New York, NY 10036

METRO BOOKS and the distinctive Metro Books logo
are trademarks of Sterling Publishing Co., Inc.

Conceived, designed, and produced by
Quid Publishing
Level 4 Sheridan House
114 Western Road
Hove BN3 1DD
England

www.quidpublishing.com

ISBN: 978-1-4351-5345-5

For information about custom editions, special sales, and premium and corporate purchases, please contact Sterling Special Sales at 800-805-5489 or specialsales@sterlingpublishing.com.

Printed in Hong Kong

1 3 5 7 9 10 8 6 4 2

www.sterlingpublishing.com

For Finn and Isaac, in anticipation of great discoveries

CONTENTS

INTRODUCTION

"If you want the exact moment in time," Kepler wrote concerning his discovery of the third law of planetary motion, "it was conceived mentally on March 8th in this year one thousand six hundred and eighteen… [before being] rejected as false, and, finally returning on the 15th of May… stormed the darkness of my mind." Among the bare facts of Kepler's account can be glimpsed the mystery and power of the instant of discovery: the blinding flash of insight that illuminates the darkness, the sudden reveal that opens a new universe to apprehension. This book recounts the extraordinary stories of 50 of history's greatest discoveries in science and exploration—eureka moments that changed the world forever, from early man literally striking the spark of inspiration in learning how to make fire to the announcement that the Higgs boson had been discovered.

How does it feel to be the first person ever to witness a natural phenomenon or set foot on alien shores? How do discoveries change the world, and their discoverers? This book explores 50 of the most important turning points in world history, detailing the excitement, joy, and awe of some of history's greatest pioneers as they make their discoveries, with gripping accounts of their adventures and breakthroughs.

In selecting the 50 episodes recounted here, I have tried to follow, however loosely, some basic guidelines. The discoveries range through history from the dawn of humanity to the modern day, and they cover many different aspects of knowledge and experience, which I have categorized under traditional subject headings: Astronomy, Biology, Chemistry, Exploration, Mathematics, Medicine, Physics, and Technology. My primary guiding principle was to focus on true "eureka" moments (named for the legendary cry of the ancient Syracusan Greek mathematician Archimedes, as he leaped naked from the bath crying, "I've got it!"—see page 24). This is to say, I have prioritized flashes of inspiration over slow dawnings, and accidental unveilings over painstaking research programs aimed at confirming expected results. On the other hand, I also wanted to include some of the great landmarks in the history of science, so I have breached my own rules in several cases, such as Darwin's discovery of evolution by natural selection, or the decades-long international effort to confirm the existence of the Higgs boson. In general, I have opted for what Isaac Newton called the *experimentum crucis* (the experiment that "points the way") rather than the working out of a theory, so the book features the 1919 Dyson–Eddington mission to observe gravitational lensing during an eclipse (see page 200), in place of Einstein's 1905 work on the theory of relativity. On the other hand, I have breached this constraint in some cases—Darwin again being one example.

What patterns emerge from reviewing the whole scope of the history of discovery? The most obvious is the increasing pace of discovery: the first five entries span all of history up to the 16th century, while by the late 19th century there are four major discoveries within the space of two years. The reasons for this acceleration are too diverse and complex to explore here, but they are almost certainly linked to the famous aphorism adopted by Newton, "If I have seen further it is by standing on the shoulders of Giants"; in other words, each discovery opens up new vistas of knowledge and brings more discoveries within reach. Also of major importance is the role of technology; while many of the earlier discoveries are appealingly accessible, in that they could have been made by anyone at any point in history, many later discoveries only became possible because of advances in technology, specifically instruments and measuring tools. The classic examples are the telescope and microscope, which made visible realms previously unavailable to human perception. Other patterns become apparent. For instance, there are three chapters in succession featuring explorers, which occupy an otherwise yawning gap in history separating important discoveries in natural philosophy. This is probably not a coincidence. Rather, it may be that the voyages of exploration were necessary precursors to the age of discovery and the dawn of the Scientific Revolution.

The 50 discoveries in this book are recounted in chronological order; if it were possible to rank them in order of importance, which would come first? The traditional answer would be Isaac Newton's discovery of the law of universal gravitation, probably vying for honors with Charles Darwin's discovery of evolution by natural selection. Both of these laws of nature have had profound theoretical and practical significance for almost all branches of science, and in diverse technologies from rockets to children's toys. Each represents the first complete revelation of a foundational element of natural philosophy, as relevant today as when it was made. But there are also strong cases to be made for other discoveries: the fire and tool-making that allowed humans to move from adapting to nature to simply adapting nature, underwriting the success of our species; or the voyages of discovery that broadened cultural vistas and, for better or worse, determined the modern economic and geopolitical shape of the world. On the basis that the most profound and momentous discoveries are those that reveal the underlying nature of reality and the systematic reality of nature, however, I would make a case that the greatest discovery of all time was that of the first mathematical laws or theorems, and so I would nominate the Pythagorean theorem (see page 19), which Kepler described as a treasure to compare with "the measure of gold."

CATEGORY

Astronomy

Biology

Chemistry

Exploration

Mathematics

Medicine

Physics

Technology

FLINT-KNAPPING: EARLY MAN MAKES ADVANCED STONE TOOLS

ca. 1.8 million years ago

Discoverer: *Homo erectus*

Circumstances: Evolution of bigger brains and key cognitive abilities facilitated step change from existing crude stone tools of the Oldowan industry

Consequences: Acheulean tool industry; global dominance of human species

[The Acheulean technology] is tremendously significant from a cognitive point of view.
I would place all this as an even more significant transition than the initial use of stone tools.

Paleoanthropologist Thomas Wynn,
University of Colorado at Colorado Springs

de Kean Collection | Staff | Getty Images

Around 1.8 million years ago, an early human—probably a member of the species known as *Homo erectus*—chose a large flint rock and struck it with another stone, at just the right angle, to produce a sharp-edged flake. Then he or she rotated the rock slightly and repeated the action; after a series of such strikes, the original rock now had a series of sharp-edged ridges terminating in a rough point, and the ground was littered with razor-sharp flakes. This was the start of a stone tool technology—or industry—known as the Acheulean, after the French town of Saint-Acheul where examples of this style of tool were first described in the 19th century by Gabriel de Mortillet. Can the advance be traced back to a single hominin (the general term for members of the species *Homo*), who was perhaps experimenting idly with the crude stone choppers made by other members of their tribe, or who was maybe even struck with a moment of inspiration while watching other apemen laboring to pierce the hide of a scavenged animal carcass? Though tempting to imagine, this "eureka" scenario is unlikely and completely unknowable. The Acheulean industry did not come from nowhere, but it did mark a great leap forward in terms of culture and technology, with profound consequences for human evolution and success.

DRAWING THE LINE

The discovery of technology is traditionally seen as the defining moment of human evolution, of greater significance than coming down from the trees or walking upright. Knowing where to draw this line proves difficult, however. Chimpanzees make extensive use of tools. They use rocks to bash other animals and crack open nuts, and use sticks to extract tasty termites from mounds; they even strip leaves from the sticks to fashion better termite extractors, in a crude form of engineering. Many other species exhibit versions of tool use, including monkeys, crows, sea otters, and octopuses.

The earliest human ancestors, such as australopithecines, could easily have been equipped from the start with chimp-like abilities to use rocks as simple tools. As far back as 2.6 million years ago these or other early hominids started what is known as the Oldowan tool industry (named for artifact finds in the Olduvai Gorge in Kenya): the first use of stone tools by early human ancestors. Oldowan tools are crude and hard to identify because they are little more than broken stones; it is probable that the industry began with opportunistic use of naturally broken tools, and that the early Oldowan tool makers learned to mimic this natural process by bashing rocks together for themselves. This seems to have been an epochal development in human history—a momentous discovery that could easily claim a chapter of its own in this book. Yet some paleoanthropologists, such as Thomas Wynn at the University of Colorado at Colorado Springs, argue that Oldowan tools are not so very different in terms of cognitive-behavioral significance from the tools used by chimpanzees: "These don't seem to represent any great intellectual leap."

The real breakthrough came when *H. erectus* started to make tools of a different order of sophistication: multipurpose, highly effective tools that required careful planning and evaluation to construct. How was *H. erectus* able to make this leap? What did the advent of the Acheulean signify? What would be the consequences of this discovery?

MASTER CRAFTERS

The tools created in the Acheulean industry were diverse but are generally characterized as bifaces and cleavers. Bifaces, commonly known as hand axes, although there is no certainty that they were used as axes, are teardrop-shaped rocks with a lenticular (lens-shaped) cross-section, formed by knocking—or knapping—flakes off a core. Cleavers are similar but have a long edge left blunt to make them easier to use for chopping. The flakes knocked off the cores may have been more useful than the cores themselves, as they could be used as knives or scrapers, and eventually perhaps even spear- or arrowheads.

© Didier Descouens | Creative Commons

BIFACES
The front, back, and sides of a typical Acheulean biface. Note the rounded "butt," which probably allowed the biface to be held comfortably in the hand.

The demands of creating such tools are many and complex. The tool-maker must be able to identify the correct type of stone and the right shape rock to use as a core, and must be able to master the difficult art of knapping. The form of the bifaces created in the Acheulean industry is striking and consistent. In some places/phases of the industry, there is even evidence that the tool-makers had in mind a specific ratio of length to width, and were thus able to appreciate proportion. Paleoanthropologist John Gowlett has shown that the dimensions of hundreds of hand axes of different sizes from the Kilombe site in Kenya, dating to around 700,000 years ago and probably made by *H. erectus*, maintained a constant proportion between length and width. In other words, whether shaping small axes or large ones, these tool-makers were able to keep in mind a "perfect" target proportion. It has even been suggested that the proportion they were aiming at was the golden ratio beloved of the ancient Greeks.

Taking the step up to this higher level of technology required the Acheulean hominins (humans and prehistoric species more closely related to humans than chimpanzees, a term not to be confused with hominids, a grouping that encompasses humans, other great apes, and their prehistoric ancestors) to be equipped with a veritable laundry list of cognitive attributes associated with the advanced brain power of modern humans. Among those attributes identified by anthropologists as necessary are large-scale spatial thinking, long-term memory, advanced planning, spatial and procedural cognition, technical and procedural know-how, communication, and social cognition. Of particular interest is the implication that communication and exchange of cultural/technological concepts must have been involved; such sharing presumably involved language. "Long-term planning and its implementation... demand an advanced form of communication," notes paleoanthropologist Naama Goren-Inbar. "From... [the] know-how and expertise shown by the high quality... and precision (symmetry and refinement) of the end-products, it is thought that the hominins [who made Acheulean tools] had language abilities."

In fact, according to a 2011 study that took casts of the inside of *H. erectus* skulls, the development of the Acheulean industry may be tied to the evolution of specific brain structures—the casts show evidence of enlargement of brain regions relevant to the mental processes involved in tool-making. So the putative apeman or -woman who first discovered that a more sophisticated bit of knapping could produce a better tool was able to do so because of major advances in brain and cognitive evolution.

Not everyone agrees that Acheulean industry is linked to a cognitive leap: the knapping process itself has been characterized as rhythmic, repetitive, and largely mindless. But Naama Goren-Inbar points to

BLOW BY BLOW

evidence from modern-day knappers in Irian Jaya, which suggests that the Acheulean industry was a dynamic, reflective process involving high-level cognitive skills such as decision-making, conversation, innovation, and flexibility: "Prolonged observations of modern-day knapping provide some insight… Knappers learn from each other and knapping sessions… are accompanied by the exchange of ideas and tips. Knapping is not simply a repeated mechanical battering but a blow-evaluation-blow procedure."

GAME CHANGER

During the Acheulean period the first evidence emerges for hunting and control of fire. Whether or to what extent these innovations can be linked to the Acheulean technological innovations is unclear, but the sophisticated stone tools that hominins were now able to produce might have led to better hunting weapons and certainly to better processing of carcasses, and hence more meat. Acheulean tools have been found in association with butchered elephant and hippopotamus skeletons, for example. This would have meant greater rewards for less effort, with increased calorie and nutrient consumption enabling the evolution of better bodies and yet bigger brains; *H. erectus* brains probably consumed 80–85 percent more energy than australopithecine ones, yet *H. erectus* was also bigger, with smaller teeth—all strongly suggesting that this species was eating much more meat.

The discovery of Acheulean tool-making brought about a new kind of evolution, known as biocultural evolution, in which humans learned to adapt to environmental change through cultural rather than biological, genetic evolution. Such adaptation can be radically faster and more successful than biological evolution, helping to explain how an essentially tropical animal such as *H. erectus* was able to spread rapidly across the entire Old World, and to do so without undergoing significant biomorphic alteration (e.g. hominins did not need to become large and blubbery to survive in Siberia or Europe; they simply found caves, built fires, and cooked meat they had hunted). Today, humans are still essentially tropical in terms of morphology, yet thrive in almost every environment on the planet.

TAMING THE FLAMES: HOW TO CONTROL FIRE

700,000–400,000 years ago

Discoverer: *Homo erectus*, Neanderthals, or archaic *Homo sapiens*?

Circumstances: Early humans' shift from opportunistic use of fire to controlling fire

Consequences: Increase in calories available from food, range of colonization, and length of day. Enables new technologies, land management, and new hunting strategies

What could be more human than the use of fire?

Harvard biologist Richard Wrangham

The two great technologies of our human ancestors were stone tools and fire; of the two, fire is the truly original discovery—no other animal is known to make even rudimentary use of fire. Along with language, fire is arguably the greatest human achievement: an accomplishment that enabled our species to colonize the world. But what does it mean to say that humans discovered fire?

FIRE MEN

There are four stages from a typical animal response to fire (stress, panic, flight) to modern mastery: understanding the concept and nature of fire; making use of fire in opportunistic fashion; control of fire; and the ability to start fires from scratch. Early hominids evolved in tropical parts of Africa, where the vast majority of the world's lightning strikes occur, and where dry seasons and active volcanoes also add to the likelihood of naturally occurring fires. Accordingly, it is likely that these hominids encountered wildfires regularly, and their reaction can be guessed at using evidence from modern-day great apes. Jill Pruetz, an Iowa State University associate professor, has studied chimpanzees in savannah conditions in Senegal and observed their relaxed and measured response to the bush fires they frequently encounter. She interprets this to mean that they have not only become accustomed to the fires, but have also built up predictive knowledge of their course and behavior, which in turn implies an ability to conceptualize fire. Pruetz links this in turn to the development of human control of fire: "If chimps can understand and predict the movement of fire, then maybe that's the thing that allowed some of the very earliest bipedal apes [human ancestors] to eventually be able to control fire."

Even so, it evidently took considerable time and the evolution of much greater brainpower for humans to progress from conceptualization of fire to use of fire, because the very earliest evidence of any kind that associated hominids with fire dates to only around a million years ago, and is widely assumed to have involved opportunistic use of naturally occurring fire (presumably in a scenario where burning material from a wildfire is picked up and used for the brief period it remains alight). This evidence comes from the Wonderwerk Cave in South Africa, where fragments of charred plants and bone (evidence of cooking?) have been found alongside stone tools dated to two million years ago, around the time of the appearance of *Homo erectus*.

What is perhaps more surprising is the lack of other evidence—not only of controlled use, but of any use—for long periods of time. At the Israeli site Gesher Benot Ya'aqov there is evidence of fire use from around 700,000 years ago, and archaeologists at the site interpret it as controlled use, but this is contentious. In a survey of evidence for early human use of fire, particularly in Europe, researchers Wil Roebroeks and Paola Villa claim that there is no evidence for controlled use of fire until 400,000–200,000 years ago, a period known as the Middle Pleistocene. In a March 2011 paper in the journal *Proceedings of the National Academy of Sciences Early Edition*, they conclude: "there was no habitual use of fire before ca. 300–400 [thousand years ago] and therefore that fire was not an essential component of the behavior of the first occupants of the northern latitudes of the Old World."

Given that there is strong evidence of hominin colonization of northern latitudes of Europe by around one million years ago, this surprising finding implies that early Europeans endured around 600,000 years of very chilly conditions. It also begs the question, why was there a gap between early opportunistic use of fire and the discovery that literally kindled human conquest of the world?

The answer lies in the complex series of tasks and challenges associated with control of fire. The ability to make fire would be a much later development, so the first challenge was to find fire. Naturally occurring fires are infrequent, impossible to predict, and dangerous. The next challenge was to keep it going, by feeding it fuel, tending it, and protecting it from the elements. This in turn meant collecting and storing fuel, keeping it dry, and stopping others from stealing flaming brands. This demanded of Middle Pleistocene humans quite advanced cognitive abilities, such as long-term planning and cooperation.

ADVANCED MAN
Control of fire may well have been part of a suite of cognitively advanced skills, such as hafting stone points to make stone-tipped spears or sewing skins to make clothes.

What was the moment of discovery? Did an observant *Homo erectus* make the link between accidental ignition of a dry bush and the possibility of gathering brushwood to make his or her own fire? Or did one triumphantly grasp a blazing brand from a wildfire and thrust it between the boughs of a tree loaded with potentially tasty prey animals? In fact, was it *H. erectus* at all? Given the evidence above that

FIREPOWER

control of fire did not come until the Middle Pleistocene, it is possible that the first hominins to control fire were the common ancestors of the Neanderthals and *Homo sapiens*, *Homo heidelbergensis* (though what these humans are called depends on the researcher), 700,000–200,000 years ago). In Europe it is likely that it was the Neanderthals who mastered the technology first (230,000–30,000 years ago).

The consequences of fire control were enormous. In his influential book *Catching Fire: How Cooking Made Us Human*, Harvard biologist Richard Wrangham argues that the most important thing about the discovery of the control of fire was that it made cooking possible, which radically boosted the accessible calorie content of the early human diet, freeing up time and energy resources for the demands of an expanding brain. In addition to cooking, fires helped humans to stay warm and survive cold seasons and environments, helped to protect them from animals, extended the hours in which they could operate, and stimulated social gathering and interaction. Fire also enabled technologies such as fire-hardening of stone and wood tools and manufacture of materials such as adhesives for hafting stone tips to wooden shafts.

IGNITION The next great discovery in the history of fire would be ignition—the means to start fires at will. There are two ways of doing this in a pre-industrial context: wood-on-wood friction or striking sparks by bashing two stones together. But both of these require specific materials: with wood friction there must be a harder wood and a softer one; with stone striking, one of the stones must be pyrites or a related iron- or sulfur-containing rock. To turn friction or sparks into a fire, highly combustible material such as dried fungus or dung is needed. The need for several elements to all be present at once and used in a highly specialized way makes the discovery of ignition technology a puzzle. One element may have been discovered through routine knapping to make stone tools, where a knapper used a pyrites-rich stone to strike a flint, producing a shower of sparks. The earliest concrete evidence for this kind of fire-making comes from the Upper Paleolithic era, with pyrites showing evidence of use as "strike-a-lights" found at Laussel in France, dating to around 25,000 years ago, although a pyrite has been found at the Grotte de la Hyene in France, a site that dates to the Mousterian era (80,000–40,000 years ago).

THE PYTHAGOREAN THEOREM: AN ANCIENT GEOMETRIC RULE

ca. 1500 BCE

CATEGORY

Astronomy

Biology

Chemistry

Exploration

Mathematics

Medicine

Physics

Technology

Discoverer: Rope-stretchers of ancient Egypt or Babylon?

Circumstances: Staking out the foundations for a structure, the master masons noticed a relationship between the lengths of the knotted strings they were using

Consequences: Monumental architecture, number theory, the science of mathematics

The day Pythagoras the famous figure found;
For which he brought the gods a sacrifice renowned!

Apollodorus the Calculator, according to Diogenes Laertius, *Lives of the Philosophers* (ca. 3rd century CE)

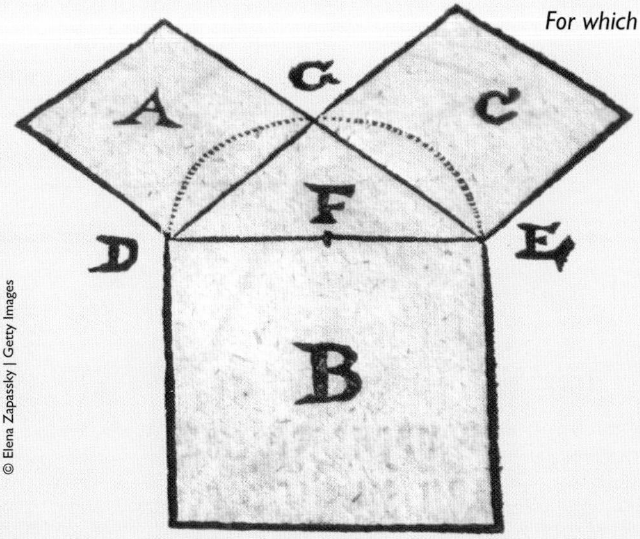

© Elena Zapassky | Getty Images

The Pythagorean theorem is a mathematical law describing the relationship between three squares, and between the three sides of a right-angled triangle. In popular legend as far back as ancient Greece its discovery was attributed to the enigmatic philosopher and mystic Pythagoras of Samos.

THE MEASURE OF GOLD

In the Greek colony of Croton, in southern Italy, an excited crowd of peculiarly dressed young men gather to watch an even more peculiarly dressed old man draw figures in the sand. He tells them of the rope-stretchers of Egypt, and the mystery encoded in the 12 knotted strings they carry, and then shows them the solution to that mystery, a revelation vouchsafed by Apollo himself. Scratching in the sand he proves to them that the squares on two sides of a right-angled triangle are equal to the square on the side opposite the angle. In her classic 1965 introduction to the history of math for children, *String, Straight-edge, and Shadow*, Julia Diggins imagines the scene: "A mighty shout—we can imagine—went up from the assembled inner group of the Secret Brotherhood… the 'mathematicians' chanted, torches waved, and smoke rose from the sacrificial altar."

In his 3rd century CE *Lives of the Philosophers*, Greco-Roman historian Diogenes Laertius relates that "Apollodorus the Calculator says that Pythgoras sacrificed a hecatomb on finding [the theorem]." A hecatomb was a ritual sacrifice of 100 oxen, made to Apollo, the god responsible for scientific inspiration; evidently Pythagoras was pleased with his discovery, and posterity would agree—Kepler described the theorem as one of geometry's great treasures, which, "we may compare to the measure of gold." But who was Pythagoras, what was his theorem, and had he really discovered it?

PYTHAGORAS OF THE GOLDEN THIGH

Pythagoras, son of Mnesarchus, came from the Greek island of Samos. Born around 565 BCE, he studied under Anaximander, who in turn had been a student of Thales of Miletus (ca. 625–547 BCE), the first named important figure in the history of mathematics. Thales was the first man recorded to have formulated mathematical theorems—provable statements in mathematics akin to natural laws in physics. For instance, one of Thales's theorems is that when two straight lines intersect, the opposing angles are equal. What makes such theorems special is that they are not simply true of the specific lines in Thales's illustration, but of any two lines ever. In other words, they are general, abstract principles. This was a new approach to mathematics that marked an exciting and momentous leap from the sophisticated but almost entirely practical and specific mathematics practiced by ancient civilizations such as the Babylonians and Egyptians.

The teachings of Thales marked the beginning of the science of geometry. According to Diogenes Laertius, "Pythagoras it was who brought geometry to perfection." As a young man eager to learn all the mathematical knowledge available in the world, he was advised to travel to Egypt, famous for the antiquity of its civilization and the breadth of its learning. According to his legend (little is known for certain about him), Pythagoras went from Egypt to Mesopotamia and Persia, and perhaps as far as India, where he may have picked up beliefs in reincarnation and unusual habits of dress (in some accounts he wore a turban). Many other myths and legends accrued to him; it was said that he had a golden thigh, could bilocate (be in two places at the same time), and could remember his past lives.

SCHOOL OF ATHENS
Pythagoras shown in a detail from Raphael's 16th-century masterpiece, *The School of Athens*. Here he appears to be studying triangular numbers and harmonic ratios rather than his eponymous theorem.

Returning to the Mediterranean, he traveled to the Greek colonies in southern Italy and set up a school in Croton. There he attracted many followers, who flocked to hear his teachings both mathematical and mystical, which included the doctrine of reincarnation and strict vegetarianism, and insisted on strange rules such as not passing an ass lying in the street, never urinating toward the sun, avoiding marriage to a woman who wears gold jewelry, and never touching black fava beans. His school became a cult, and it is from the initiates of this Pythagorean Brotherhood that the word mathematician derives. New members signed over all their possessions and took a vow of silence for five years, during which time they were known as *akousmatikoi*, or listeners, allowed to follow Pythagoras's lectures from behind a curtain. Eventually they could become *mathematikoi* ("those who learn"), initiates of the inner circle. For the cult, mathematics was the route to transcendence. According to 5th-century CE philosopher and historian Proclus, "among the Pythagoreans there was prevalent this motto, 'A figure and a platform, not a figure and sixpence,' by which they implied that the geometry deserving study is that which, at each theorem, sets up a platform for further ascent and lifts the soul on high, instead of allowing it to… fulfill the common needs of mortal men…"

Among the theorems that best exemplified this mission, that which is now known as the Pythagorean theorem ranked high. The theorem says that, for a right-angled triangle, the square on the hypotenuse is equal in area to the sum of the squares on the other two sides. So if the length of the hypotenuse is a, and the lengths of the other two sides are b and c, the theorem expressed in algebraic terms (not to be invented for over a thousand years after the time of Pythagoras) is $a^2 = b^2 + c^2$.

THE ROPE STRETCHERS

Supposedly, Pythagoras was inspired to develop his theorem after encountering while in Egypt men known as "rope-stretchers"— presumably surveyors. They were said to use an ancient technique involving a rope divided by knots into 12 equal units of length. If they staked out one end, counted four units, staked the rope again, counted out five units and then staked it such that the remaining three units of rope reached back to the start of the rope, they could create a right-angled triangle, an essential tool for surveying (being able to set right angles was a precondition for monumental masonry on the scale of the pyramids). The triangle thus created would have sides of 3, 4 and 5 units; plugging these values into the formula above reveals that $3^2 + 4^2$ does indeed equal 5^2. In fact, what the Egyptian rope-stretchers had stumbled upon was the simplest form of what are known as Pythagorean triples: sets of three whole numbers that fit the Pythagorean equation, which is to say, right-angled triangles where the lengths of all three sides are whole numbers. For instance, the next Pythagorean triple is 5, 12, and 13.

The ancient Egyptians must then have already known about the Pythagorean theorem, at least in rudimentary form. They weren't the only ones. The ancient Babylonians had also known the theorem, since at least 1900 BCE. The best evidence for this is an ancient cuneiform tablet known as Plimpton 322 (it comes from the Plimpton Collection at Columbia), which dates to the Old Babylonian Empire, ca. 1800 BCE. The tablet includes a table of values that appears to be a sort of list of Pythagorean

PLIMPTON 322
The famous cuneiform tablet known as Plimpton 322, which has been contentiously interpreted as indicating a knowledge of Pythagorean triples.

triples, according to pioneering archaeologist Otto Neugebauer. This is not universally accepted, but there is other evidence that the Babylonians knew the principle of the Pythagorean theorem. Another Old Babylonian Empire tablet preserved in the British Library includes calculations for the squares of the sides of a 3, 4, 5 right-angled triangle, reading, "4 is the length and 5 the diagonal. What is the breadth?" The ancient Babylonians may also have been familiar with the simplest and easiest of all demonstrations of the theorem, known as the tile proof because it can be seen in a tile pattern that may have been known to the ancient Babylonians. The ancient Chinese and Indians also knew of the theorem; an illustrated treatment is found in the oldest surviving Chinese mathematical text, the *Arithmetic Classic of the Gnoman and the Circular Paths of Heaven*.

In other words, Pythagoras did not discover the principle of the theorem that bears his name; possibly it was first perceived by rope-stretching surveyors at work in the very earliest civilizations of Mesopotamia, Egypt, or China. But Pythagoras may have been the first to make the leap from the very specific instances known to the ancients to the much more profound understanding encapsulated in the Pythagorean theorem. Pythagoras might also have been the first to prove the theorem—an important step that marks the discovery that an interesting potential relationship between numbers and lines is an immutable law of the universe. Proving the theorem opened the door to a new world of mathematics, and by extension to entire realms of philosophy and science that underpin all of modern life.

THE GEOMETRY DESERVING STUDY IS THAT WHICH, AT EACH THEOREM, SETS UP A PLATFORM FOR FURTHER ASCENT AND LIFTS THE SOUL ON HIGH, INSTEAD OF ALLOWING IT TO FULFILL THE COMMON NEEDS OF MORTAL MEN.

CATEGORY

Astronomy

Biology

Chemistry

Exploration

Mathematics

Medicine

Physics

Technology

THE ARCHIMEDES'S PRINCIPLE: THE ORIGINAL "EUREKA!" MOMENT

ca. 230 B CE

Discoverer: Archimedes of Syracuse

Circumstances: Musing on how to test the purity of a golden crown by means of determining its volume, Archimedes took a bath…

Consequences: Principle of buoyancy, science of hydrostatics

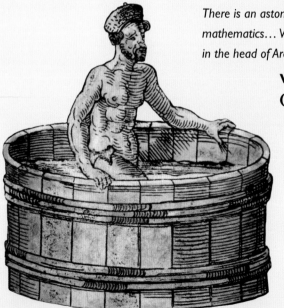

There is an astonishing imagination, even in the science of mathematics… We repeat, there was far more imagination in the head of Archimedes than in that of Homer.

Voltaire, *Philosophical Dictionary* (1764)

The original "eureka" moment in history marked the discovery by Archimedes of Syracuse of the principle of buoyancy, which states that the buoyant force is equal to the weight of the displaced water (i.e. something will float if the water it displaces weighs more than it does). This was just one of Archimedes's discoveries and achievements; he also discovered the law of the lever, famously demonstrating its power by single-handedly dragging a heavily laden galley across the ground.

The tale of Archimedes in the bath tub is the climax of a story related by the ancient Roman writer Vitruvius, in his 1st century BCE *On Architecture*, about efforts to prove whether an artifact made for the king of Syracuse had been made of adulterated gold. In the tale, Hiero, the king, gave a nugget of pure gold to a smith to fashion a beautiful crown or laurel wreath, but after receiving the cunningly fashioned diadem he suspected the craftsman of having cheated him by adulterating the gold with a lesser metal such as silver (and selling the leftover gold). The crown weighed the same as the original raw material, but this proved nothing. Hiero asked Archimedes — "a kinsman and friend," according to the Roman writer Plutarch — for help.

THE BOGUS CROWN AFFAIR

Archimedes knew that since gold was the densest metal known in the ancient world, if some of the gold in the crown had been replaced by silver or tin, the mixture would be less dense than pure gold, so that although the original nugget and the crown might weigh the same, if the crown were not pure it must therefore have a greater volume than the original nugget. Unfortunately, for such an intricate and complex shape as a laurel wreath, there was no way to measure its volume. Mulling over the problem, Vitruvius recounts, Archimedes went to the baths and settled into the tub: "as he was sitting down… he noticed that the amount of water which flowed over the tub was equal to the amount by which his body was immersed." In other words, his body displaced an amount of water equal to its volume. This must be true for any body/ object, and Archimedes realized that he had discovered a simple but ingenious way to measure the volume of any object, no matter how intricate. "In his joy," Vitruvius relates, "[he] leaped out of the tub and, rushing naked toward his home, he cried with a loud voice that he had

GEOMETER AT WORK
Domenico Fetti's 17th-century depiction of Archimedes shows him with the tools of the geometer's trade, the compass and straightedge.

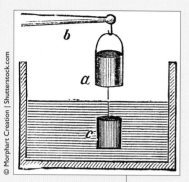

TESTING THE PRINCIPLE

A device to verify Archimedes's principle by experiment, using a solid cylinder and a bucket of precisely the same volume, both suspended from a spring balance.

found what he sought. For as he ran he repeatedly shouted in Greek, *eureka, eureka.*" "Eureka" is Greek for "I have it!"

According to the legendary tale, Archimedes solved the mystery by placing into a bowl a nugget of pure gold of equal weight to the crown, and then filling the bowl to the brim with water. He then swapped the nugget for the crown itself, and water slopped over the sides, proving that the crown had a greater volume than an equivalent mass of pure gold, and thus a lower density. In other words, it had indeed been adulterated. Unmasked, the crooked craftsman was sent to a suitable doom.

Galileo and many others have pointed out that in practice the "nugget in a bowl" method would not offer the resolution needed to detect the tiny difference in volumes between the nugget and the crown. A more likely solution is that Archimedes used an ingenious combination of his discoveries about buoyancy and levers to design a simple balance-in-water. In this method, the nugget and the crown are suspended from either end of a balance; being the same weight, they balance out and the balance remains horizontal. Once immersed in water, the less dense, more buoyant crown displaces more water and hence experiences a greater force of buoyancy pushing it upward, so it rises and the balance tilts.

THE LAW OF THE LEVER

"Give me a place to stand and I will move the Earth." This famous remark is attributed to Archimedes by the ancient mathematician Pappus of Alexandria in his *Synagoge* of ca. 340 CE. It refers to Archimedes's discovery of the Law of the Lever; of greater significance than his bathtime breakthrough in buoyancy, though less memorably celebrated. In Book I of his *On the Equilibrium of Planes*, Archimedes states the Law as "Magnitudes are in equilibrium at distances reciprocally proportional to their weights." What this means in practical terms is that to find the force exerted on one end of a balance or lever, the weight pressing down is multiplied by the distance from the fulcrum (the point on which the balance rests or the lever hinges).

For instance, Jane, who weighs 7 stone (45 kilograms), sits on one side of a balance while her brother John, who weighs 14 stone (90 kilograms), sits on the other side. If John sits 1 meter (3.3 feet) from the fulcrum, how far from the fulcrum should Jane sit in order to balance him? The force exerted by John is 14 x 1 = 14 stone/m. Jane needs to sit 2 meters from the fulcrum to generate the same force (7 × 2 = 14); simply by sitting farther away from the fulcrum she can balance her much heavier brother. The lever (in this case a balance beam resting on a fulcrum) is the simplest form of machine, which is to say, a device for doing work. At the point where Jane is sitting she exerts a downward force (known as the input force) of just 7 stone, but the lever converts this into an upward force (known as the output) of 14 stone acting on John at the other end. The ratio between the output and input force is known as the mechanical advantage; in this case, the balance beam gives a mechanical advantage of 14/7, or 2.

The term "lever" typically conjures up the image of a simple bar or stick, but levers actually come in many forms. A wheel is a type of lever—a small force exerted on the outer rim gives a large force at the axle, just as a small force applied at the end of a wrench or spanner becomes a great force acting on the nut being tightened or loosened. Another type of lever is a pulley, where a rope runs round a wheel or combination of wheels.

WITHOUT ANY SPECIAL EFFORT

Archimedes used a pulley to accomplish another of his famous feats. In his work *Marcellus*, the Greco-Roman writer Plutarch tells a story of how Archimedes wrote to King Hiero to tell him how he had proved the Law of the Lever, leading to his famous boast that "if there were another world and he could go to it, he would move this one," as Plutarch puts it. Hiero asked Archimedes "to give a practical demonstration of the problem," and the philosopher "chose a three-masted merchantman among the king's ships which had been hauled ashore with great labor by a large band of men," and ordered it loaded with men and cargo. Then he attached a compound pulley to the ship and, "without any special effort, he pulled gently with his hand… and drew the vessel smoothly and evenly toward himself as though she were running along the surface of the water."

IRON CLAW
One of the most striking applications of Archimedes's mechanical genius—and particularly his expertise with levers—was his infamous "iron hand" ship claw, a device for overturning attacking ships.

CATEGORY

Astronomy

Biology

Chemistry

Exploration

Mathematics

Medicine

Physics

Technology

AN EXPLOSIVE REACTION: THE SECRET OF GUNPOWDER

ca. 850 CE

Discoverer: Tang dynasty alchemist

Circumstances: Experimenting with elixirs to prolong life, Chinese alchemists tried various combinations and proportions of charcoal, saltpeter, and sulfur

Consequences: Revolutionized warfare and inspired the invention of rocketry

Some have heated together the saltpeter, sulfur, and carbon of charcoal with honey; smoke and flames result…

Taoist alchemical text (ca. 850 CE)

Poring over their ancient text, the alchemists muttered sagely to one another and nodded at their racks of vials and bottles. They had all the ingredients necessary: black charcoal from burned willow, yellow sulfur, and white saltpeter crystallized from distillation of guano. Wei Boyang described the transformative effects of heating sulfur and saltpeter; now they would try adding charcoal in order to achieve the ultimate goal—the elixir of immortality. Onto the brazier over the little fire they poured the charcoal, stirring in the sulfur, and then tossing in the saltpeter. The result was an explosion heard around the world, which would reverberate down through history.

Whether this accurately describes the scene that attended the discovery of gunpowder is impossible to say, but according to the first recorded account of gunpowder (aka black powder), preserved in a Chinese Taoist alchemical text of ca. 850 CE, "some have heated together the saltpeter, sulfur, and carbon of charcoal with honey; smoke and flames result, so that their hands and faces have been burned, and even the whole house burned down." These accidental explosives engineers were building on a long tradition of Chinese alchemy, said to go back at least as far as the legendary First Emperor, Qin Shi Huang (260– 210 BCE), who pursued a desperate search for the elixir of immortality. This continued to be an obsession of Chinese alchemy, and sulfur and saltpeter (potassium nitrate, obtained in ancient times from guano or urine) were both tested in this respect and (presumably) found wanting. Alchemists experimented with heating the substances in their quest to achieve transmutations, and sulfur and saltpeter were among the substances described in the earliest alchemical text known from China, Wei Boyang's 2nd-century *Book of the Kinship of the Three*. This was the context in which the 9th-century alchemists made their explosive discovery, but its military applications were not immediately apparent and early uses included insect fumigant and skin-disease treatment.

The black powder the alchemists had created combined all the elements needed for an explosive, which is a substance that burns so quickly that it generates great amounts of heat in a very short time while turning solid reactants into gaseous products. The result is very rapidly expanding hot gas, with the potential to impart a great deal of kinetic energy to anything in close proximity. In gunpowder the fuel is carbon

BURNING DOWN THE HOUSE

in the form of charcoal, traditionally obtained by burning willow, but which has also been made from grapevine, hazel, elder, laurel, and pine cones; sugar is also used.

In the explosive reaction the carbon burns to give carbon dioxide, and in order for it to burn quickly enough to be explosive requires a reactant that can supply large amounts of oxygen very quickly—an oxidizer. This is where the saltpeter, or potassium nitrate, comes in. A powerful oxidizer, it can make almost any organic material burn fast; in his 1643 *Mysteries of nature and art*, John Bate, a maker of gunners' scales, described it as "the Soule of Gunpowder."

More mysterious is the role of sulfur, which is still poorly understood and often misrepresented. The gunpowder mixture will burn without sulfur, but far less effectively. Sulfur burns at a much lower temperature than the charcoal and saltpeter, lowering the ignition point of the mixture. Once it is burning it instantly raises the temperature of the mixture to the fusion point of saltpeter, setting off the explosive reaction, and it also acts to increase the speed of combustion. Without sulfur, wrote master gunsmith Biringuccio in his 1540 *Pirotechnia*, gunpowder "would be nothing, because it would be impossible to introduce the fire instantaneously throughout the powder so that it will ignite as it is seen to do."

By the Song dynasty, the Chinese had incorporated gunpowder into their military stratagems and tactics. The earliest genuine recipe for gunpowder comes from the 1043 manual *The Essentials of the Military Arts*, which gives instructions and illustrations for a range of devices, including firecrackers, bombs, bamboo rockets, and gunpowder-rocket-assisted fire arrows. According to Robin Yates, Professor of History and East Asian Studies at McGill University, "Although scholars often consider the Song Dynasty to have been very weak, its use of gunpowder was the reason it was able to hold off the Mongols for many decades."

The Song Chinese tried to protect the secret of gunpowder, but eventually the Mongols learned it and developed their own gunpowder weapons. Through them knowledge of the revolutionary discovery spread along the Silk Road and was transmitted to Europe via the Islamic world. There are claims that Europeans knew the formula for gunpowder

© Steve Sstvanik | Shutterstock.com

as early as the Chinese; for instance, the *Liber Ignium ad Comburendos Hostes* (Book of Fire for the Burning of Enemies), attributed to Marcus the Greek, includes a recipe claiming to date to the mid-9th century, but in fact it is probably a late addition to the text, which was most likely compiled around 1300. Probably the earliest reference to gunpowder in Europe comes from around 1260, when the English friar Roger Bacon described what were almost certainly firecrackers; he is also often, though contentiously, credited with the first European recipe for gunpowder.

Once gunpowder reached Europe it began to transform warfare and society with almost alchemical power. Slowly at first, but with ever increasing impact, gunpowder upended the military status quo and then the social and political order. Initially, gunpowder weapons technology was too crude for any use other than short-range, indiscriminate blasting, but this was exactly what was needed for siege warfare. Great bombards of brass and iron could blast holes in even the mightiest of medieval fortifications, and monumental stone walls that had dominated strategy and socio-politics for centuries (or in the case of Constantinople, millennia) crumbled beneath the onslaught. Perhaps the most symbolic example was the Fall of Constantinople in 1453, where the Ottoman sultan Mohammed II employed 56 cannon and 12 great bombards, including one monster known as Basilica, which took an hour to load and fired a ball weighing 114 stone (725 kilograms) just under a mile (a kilometer and a half). Gunpowder succeeded where generations of foes had failed and the mighty city walls were finally breached.

Four years earlier the French king, Charles VII, had taken 60 castles and fortified towns in the course of 16 months thanks to his cannon.

FALL OF CONSTANTINOPLE
Diorama from a museum in Istanbul, showing the final assault on the wrecked walls of Constantinople, with colossal bombards in the foreground.

GUNPOWDER CHANGES THE WORLD

MONGOL WARRIORS
Japanese painting from
the 13th century showing
Mongol warriors using
gunpowder weapons; it
was probably via them that
gunpowder spread west
from China.

Gunpowder rendered the castle obsolete, removing the stronghold of the feudal lord and tipping the balance of power toward the centralizing state, thus playing a decisive role in the development of the nation state and the collapse of the old feudal order.

On the battlefield too, gunpowder began to change things. Even the early, highly immobile cannon chalked up some significant victories such as the final French defeat of the English in the Hundred Years War. Portable firearms in the form of muskets became increasingly common, proving to be a decisive leveler in the long struggle between heavily armored aristocrats, usually on horseback, and the infantry, too poor to afford armor and not previously equipped with the means to meet the knights on equal terms. With a musket ball propelled by a gunpowder blast, however, the meanest musketeer could pierce the armor of a king: gunpowder signaled the end of feudal tactics on the battlefield and would eventually render the combat horse obsolete. The ever increasing demand for gunpowder would work more subtle changes on the battlefield, as the need for reliable supplies of good quality powder resulted in the development of the science of logistics, heralding the eventual full-scale industrialization of war.

The discovery of gunpowder would have technological consequences beyond the military. Gunpowder had proved early on to be an excellent propellant for rockets, and would play an important role in the early experiments of interwar rocket scientists in America and elsewhere, experiments that would later blossom into the space program. Even the multiple stage rocket design that would be needed for the moon shot was prefigured in early Chinese gunpowder weaponry. The idea of explosive expansion of gases in a tube acting as a motive force—as achieved by early Chinese gunpowder weapon developers—has even been credited as the inspiration for the internal combustion engine.

THE NEW WORLD: COLUMBUS SAILS THE OCEAN BLUE

1492

CATEGORY

Astronomy

Biology

Chemistry

Exploration

Mathematics

Medicine

Physics

Technology

Discoverer: Columbus (et al)

Circumstances: Seeking a new and lucrative route to the Indies, Columbus sails west from Europe. However, he is not the first to have done so…

Consequences: Opening of the Americas to European colonization

To Castille and to Leon, A NEW WORLD gave Colon.

Epigraph on tomb of Christopher Columbus (aka Cristobal Colon)

FAMED DISCOVERER
A painting from the
Metropolitan Museum in
New York generally held
to be the authoritative
depiction of Columbus.

MAN WITH A PLAN

On the evening of Thursday, October 11th, 1492, Christopher Columbus was in a state of anticipation, excitement, and anxiety. The rumblings of his crew, driven mutinous by long weeks of sailing through trackless ocean, had been quieted by signs of imminent landfall: land birds, floating vegetation, and flotsam. "Everyone breathed afresh and rejoiced at these signs," Columbus recorded in his journal. Staring westward, straining his eyes to peer through the dark, he saw a flash of light, "like a wax candle rising and falling." He must have been flooded with relief—after a month of sailing and thousands of nautical miles, he had been proven right: it was possible to reach the Indies by sailing westward! Of course, Columbus was mistaken, about a great many things. He had not reached the Indies (in the sense of China, Japan, and South East Asia), but the Caribbean, although his error is preserved in the name given to the islands of that sea, the West Indies. But today he is celebrated as the first man to discover the New World of the Americas, and 1492 marks a turning point in world history.

Born Cristoforo Colombo in Genoa in 1446, Columbus, as he is known today in the English-speaking world, was the son of a weaver, but his family seems to have caught something of the brewing excitement in Renaissance Europe about new worlds awaiting discovery. His brother would go on to become a map-maker, while Christopher trained as a seaman and dreamed of exploration. He devoured all the information he could find about distant lands, mixing relatively recent revelations, such as Marco Polo's sensational accounts of Cathay and Cipango (China and Japan), with antique inspirations such as Plato's Atlantis. Rumors swirled of legendary islands in the Atlantic—Hy Brasil, the Seven Cities of Antilia, the Happy Isles of St. Brendan. Meanwhile, the Portuguese, having laboriously explored the coast of Africa (see page 38), were on the verge of proving that Europeans could reach and exploit rich lands on the other side of the world by a maritime route, opening the prospect of breaking the Venetian-Islamic monopoly on trade with the Far East.

In 1473, Columbus and his brother settled in Portugal, the European center for exploration, and the following year he heard of a new map drawn up by an Italian philosopher, Toscanelli, which combined the ancient geography of Ptolemy with information from Marco Polo. Cathay, Toscanelli concluded, lay as little as 3,120 miles (5,020 kilometers) from the Azores, the most westerly lands known to the Europeans. The sea route to the Indies lay open to the west, if only men were bold enough to strike out into the unknown. This at least was how it appeared to Columbus, who adopted Toscanelli's map as his guide. Like all educated people at this time, they knew the world to be spherical—they just failed to appreciate how big a sphere it is.

First Columbus approached the king of Portugal with his plan for an expedition to the west, but he was rejected. According to legend the Portuguese cravenly attempted to steal Columbus's plan and sent, in secret, an unnamed sailor to attempt the crossing, but the man took fright on losing sight of land and turned back. In 1487 Columbus first approached the Spanish monarchs, Ferdinand and Isabella, but it was not until 1492 that the project finally secured the right backing and the Genoese was commissioned as Admiral of a fleet of three ships (named, or nicknamed, the *Santa Maria*, *Nina*, and *Pinta*) and Viceroy of all the lands he might discover. He set sail from Spain on August 3rd, spent several weeks in the Canaries, and eventually struck out into the open ocean on September 6th, steering due west.

Did his crew know they were engaged in a leap of faith? Certainly, Columbus doubted their commitment and resolve, for he started to keep two records of the little fleet's progress—an accurate one for his royal patrons, and one for his crew that showed a much reduced estimate of the distance covered "that the men might not be terrified if they should be long upon the voyage." By October 10th, however, "the men lost all patience, and complained of the length of the voyage," but Columbus mollified them with talk of "the profits they were about to acquire," and pointed out that "it was to no purpose to complain, having come so far, they had nothing to do but continue on to the Indies." The next day came the momentous sighting of a light, and in the morning of the 12th a sailor called Rodrigo de Triana was the first to spy land. People were spotted, and Columbus went ashore bearing the standards of Spain and Christendom. "Arrived on shore, they saw trees very green many streams of water, and diverse sorts of fruits," he noted, although it was the small gold discs worn on the ears of many of the natives that particularly grabbed his attention.

IT WAS TO NO PURPOSE TO COMPLAIN, HAVING COME SO FAR, THEY HAD NOTHING TO DO BUT CONTINUE ON TO THE INDIES.

The true purpose of Columbus's expedition was not exploration, however, but conquest and profit. For the unfortunate natives this was the beginning of a holocaust. Disease, slavery, and mistreatment would wipe out entire populations and civilizations; Columbus himself proved to be brutal and unjust in his dealings with the indigenous peoples. Yet he is celebrated to this day with a national holiday in the USA and many other countries, and he is still popularly regarded as the man who "discovered America." In practice, Columbus never set foot on North America, although on his last voyage he did reach mainland Central America. He went to his grave convinced that he had discovered the Indies and that Cathay and Cipango lay near at hand.

FIRST DISCOVERERS

More significant is the question of the extent to which America can be said to have been discovered in 1492. By this time the original humans to set foot on the continents of North and South America had been there for at least 14,000 years, which is the date at which a group of humans known as the Clovis peoples or culture (named for the New Mexico site where their distinctive spear points were first found in 1932) spread rapidly from Alaska to Patagonia.

Columbus was not even the first European to discover the Americas. Vikings under Erik the Red had settled Greenland in the 980s, and in 985 a ship making the passage there was blown off course and sighted land to the west. Erik's son Leif set off to find this land around 1000, reaching what is now Newfoundland, which he called Vinland. It is possible that Basque and Breton fishermen strayed that far in pre-Columbian times; they were certainly among the first to exploit the fisheries off the North American coast after John Cabot became the first European since the Vikings definitely to have made landfall in North America, in 1497. More plausible is the possibility that Polynesian seafarers reached the western coasts of the Americas before Europeans; pre-Columbian chicken bones from Chile seem to have come from Polynesia, while the sweet potato, which originated in South America, may have made the journey the other way. Wilder candidates for pre-Columbian explorers include ocean-going medieval Irish monks, Ming Dynasty Chinese fleets, Roman and/or Phoenician merchants, and ancient Egyptians.

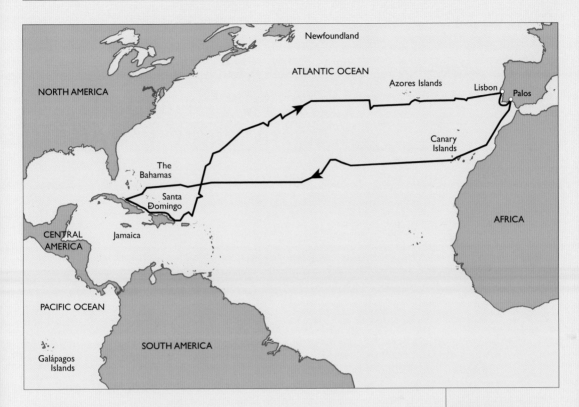

If Columbus was not the man who discovered America, what did he achieve? From a contemporary European perspective his achievements were profound. In practical terms he proved that the northeast trade winds offered a feasible—even convenient—passage across the Atlantic, while the westerlies would carry ships home to Europe; this had not been known before, certainly not by Columbus at any rate. In one voyage he had laid down the basic itinerary for what would become a trans-oceanic "super-highway." Less tangibly he shattered whatever psychological "invisible barrier" may have been blocking European mariners from contemplating transatlantic voyaging. In geopolitical terms he broke the Portuguese monopoly on trans-oceanic trade and colonization, and set Spain on a rollercoaster ride to European supremacy (however temporary). The opening of the New World shifted the focus of Europe westward, and utterly changed the destiny of the Americas. Whether or not he discovered America, Columbus's landfall shook the world.

TRANSATLANTIC ROUTE
The first voyage of Christopher Columbus. Perhaps his most significant achievement was to establish the best routes for transatlantic sailing using the prevailing winds at each latitude.

CATEGORY

Astronomy

Biology

Chemistry

Exploration

Mathematics

Medicine

Physics

Technology

BY SEA TO THE INDIES: DA GAMA'S EPIC VOYAGE

1497

Discoverer: Vasco da Gama

Circumstances: The king of Portugal commissioned Vasco da Gama to travel to India

Consequences: Opened the maritime route to the Indies, establishing decades of Portuguese dominance of trade; seeded a string of Portuguese colonies

May the Devil take thee! What brought you hither?

Tunisian Moors on meeting da Gama in India, 1497, as recorded in Vasco da Gama's journal

A few years after Columbus had failed to reach the Indies by the western sea route, the Portuguese mariner Vasco da Gama completed the eastern route, in the dramatic culmination of a 60-year long program of incremental exploration. His journey was difficult, dangerous, and bloody, it ended in humiliation and hasty retreat, and it brought back a relatively small amount of booty, yet it was one of the most successful and important voyages of the entire Age of Exploration.

Portugal, with its long Atlantic coast and poverty of natural resources, perhaps inevitably looked to the sea to make its fortune. It would go on to become one of the richest and most powerful nations on Earth for a period. First, however, medieval sailors had to overcome a mixture of superstitions and quite reasonable fears. Lack of navigation aids meant that sailing out of sight of land was dangerous and frightening. The Atlantic to the southwest of Portugal was stormy, while according to legend the sea near the Equator was boiling hot and infested with monsters. It was generally accepted that no one who had ventured below latitude 29° north had ever returned.

In the 15th century, however, one of the key figures of the Renaissance emerged. Prince Henry the Navigator (1394–1460), as he came to be called despite doing no seafaring of his own, was the son of John I of Portugal. His horoscope specifically predicted that he would make important discoveries, and he dreamed of opening a route to the Orient for a mixture of economic, political, and religious motives. The lucrative trade with the Indies for spices and other luxury goods was controlled by the Ottomans and their trading partners in the Italian city states, primarily the Venetians. If there were a sea route that bypassed the Silk Road and the Muslim-controlled lands of the Middle and Near East, it could prove extraordinarily valuable. At the same time, there was a widespread belief in the existence of Christian kingdoms and kings in the Orient, such as the legendary Prester John, who could prove powerful and strategically vital allies against the Moors (the generic term used to refer to Muslims at the time). For those far-off lands that were not Christian, there was an opportunity to spread the faith by converting benighted pagans.

Henry set up a school of navigation and seamanship, and chartered a series of ever-more ambitious voyages to explore the coastline of Africa,

PRINCE HENRY'S DOOM

having started by settling the Madeira Islands in 1420. In 1434 Gil Eannes passed the "father of danger," as the Arabs called the dangerous shoals off the West African coast, and became the first European known to have rounded Cape Bojador (in what is now Western Sahara). In 1441 Nuno Tristão reached Cape Blanc; in 1444 Dinis Dias discovered the Cape Verde Islands (off what is now Senegal); and in 1446 Tristão reached the mouth of the Gambia River.

The Spanish began sending their own expeditions to Africa, raising the stakes, and in 1487 Bartholomeu Dias, venturing far to the south, was caught up in a storm and blown past what later came to be called the Cape of Good Hope—the southern tip of Africa. It was now clear that it was possible to sail round Africa. This important discovery was given added value by the intelligence gathered by the intrepid Portuguese spy, Pedro da Covilha, a fluent Arabic speaker who was dispatched on a dangerous overland mission to scout East African and Indian ports.

In 1492 Columbus returned with claims of having pioneered a western route to the Indies. Portuguese expeditions to West Africa had garnered some wealth through gold and slavery, but the real wealth lay in the Indies. The pressure was on for Portugal to make its long program of exploration pay off. In 1495 a new king, Manuel I, came to the Portuguese throne, and he soon started work on planning an expedition to the far side of Africa. To helm the little squadron that would be sent out he needed a suitable leader, and settled on Vasco da Gama, son of a nobleman who had commanded the fortress at Sines, and a proven mariner who had previously commanded punitive raids on French shipping.

ZACUTO'S CHARTS

The expedition was planned and prepared meticulously. It would be armed with the intelligence gathered by Covilha and the experience of Diaz, who personally oversaw the preparations. Da Gama would command four ships, two of them specially constructed to meet the challenges ahead: *naos*—square-rigged sailing ships of 200 tons. The *St. Gabriel* would be da Gama's flagship, while the *St. Raphael* would be captained by his brother Paolo. In addition, there was a lateen-rigged caravel under Berrio Nicolau Coelho and a store ship. Among the 170 crew of the squadron were three experienced pilots, including one who had sailed with Diaz.

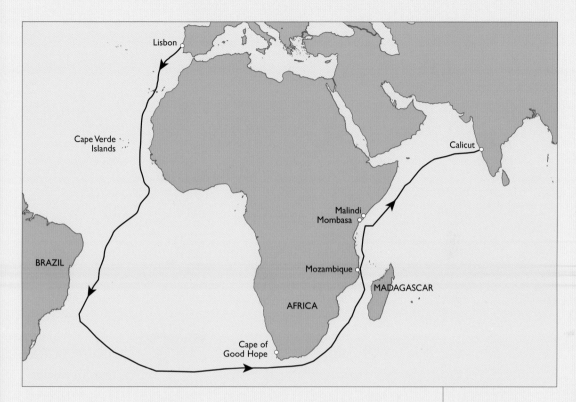

Da Gama also shared an important advantage with Columbus: charts and instruments from Abraham Zacuto, the foremost Spanish scientist of his day and a major force in the emerging science of cartography. Zacuto had published an important set of astronomical tables that Columbus had carried with him to the New World, and Zacuto's eminence was such that in 1504 he was able to boast, "My astronomical charts circulate throughout all the Christian and even Muslim lands." Unfortunately, this was partly because he was successively chased out of three kingdoms by anti-Semitic Iberian crusades. In 1492 he was expelled from Spain with the rest of the Jews and settled in Lisbon, where he became court astronomer and historiographer to John II and his successor Manuel. Consulted in the run-up to da Gama's expedition, he offered support and encouragement, as well as practical help in the shape of an iron astrolabe (a navigational instrument) of his own design. Despite these services to the crown, Zacuto and his family would be driven out of Portugal and later chased across Africa to take refuge with the Ottomans.

A LONG HAUL
Vasco da Gama's epic journey, including the wide westward swing across the Atlantic, necessary to avoid the doldrums, a zone of lackluster winds, which brought him within a few days' sailing of Brazil.

AROUND THE HORN

DECORATED COMMANDER
Portrait of da Gama wearing the cross of a Commander of the Portuguese Military Order of Christ, awarded to him by King Emmanuel I in 1507, following his second return from India.

On July 8th, 1497, da Gama's squadron set off from Lisbon. Diaz himself piloted them as far as the Cape Verde Islands, and they followed his advice to avoid the doldrums, taking a huge swing far out into the western Atlantic, which brought them close to Brazil, before catching the south-westerlies back to Africa. After 13 weeks out of sight of land they made landfall just to the north of the Cape of Good Hope, and rounded the Cape just before Christmas. Over the next few months they worked their way up the east coast of Africa, but found scant welcome at the trading ports of Mozambique or Mombasa. The ruler of Malindi proved more friendly and provided them with a pilot to take them across to India, and on May 20th, 1498 da Gama reached Calicut (now Kozhikode).

His initial contacts were not encouraging, for the first two people he met were "two Moors from Tunis, who could speak Castilian and Genoese." In his Journal, da Gama records that "The first greeting that he received was in these words: 'May the Devil take thee! What brought you hither?' They asked what he sought so far away from home, and he told them that we came in search of Christians and of spices." Indeed, the Portuguese mistook the local Hindus for Christians.

In his first interview with the local potentate, da Gama laid out the background to his voyage of exploration: "… that for a period of 60 years [the King of Portugal's] ancestors had annually sent out vessels to make discoveries in the direction of India… [but] the captains sent out traveled for a year or two, until their provisions were exhausted, and then returned to Portugal, without having succeeded in making the desired discovery… Dom Manuel… had ordered him not to return to Portugal until he should have discovered this King of the Christians, on pain of having his head cut off." He was, however, less than frank as to the real purpose behind such expeditions, claiming that the Portuguese were seeking to reach India "as they knew that there were Christian kings there like themselves. This, he said, was the reason which induced them to order this country to be discovered, not because they sought for gold or silver, for of this they had such abundance that they needed not what was to be found in this country."

In fact, da Gama was desperate to secure lucrative trade and even more lucrative trading rights, but the paltry gifts he had to offer were spurned and his stay began to turn sour. After being placed under armed guard for a period, da Gama was finally allowed to leave, returning to Portugal in July 1499, thoroughly depressed at the death of his brother en route. After more than two years away from home, including 300 days at sea, da Gama had covered 24,000 miles (39,000 kilometers), lost two ships and all but 54 of his original crew.

Dom Manuel immediately dispatched a follow-up expedition, but when this ended in bloodshed, da Gama was sent back to India in 1502 at the head of a fleet. He destroyed much of Calicut and tortured and executed many Indians, returning home laden with booty. A third and final trip to India in 1524 was short-lived, as he died soon after arrival.

The initial voyage of Vasco da Gama blazed a trail that would become well-trodden in the years that followed; specifically his loop westward would turn into a triangular itinerary that saw gold, slaves, and sugar flow back and forth across the Atlantic. Portugal would go on to establish lucrative "factories" (trading colonies) along the coasts of Africa and in India and the Far East, with incredible wealth flowing into the coffers of the Manuel dynasty. His voyage of exploration had an impact beyond the purely economic, however, broadening European horizons and ushering in a new era of intellectual as well as mercantile exchange.

DOM MANUEL HAD ORDERED HIM NOT TO RETURN TO PORTUGAL UNTIL HE SHOULD HAVE DISCOVERED THIS KING OF THE CHRISTIANS, ON PAIN OF HAVING HIS HEAD CUT OFF.

CATEGORY

Astronomy

Biology

Chemistry

Exploration

Mathematics

Medicine

Physics

Technology

INTO THE JUNGLE: ORELLANA DISCOVERS THE AMAZON

1541–42

Discoverer: Francisco de Orellana

Circumstances: Part of a disastrous conquistador expedition across the Andes; Orellana and 50 men were sent downriver to search for food

Consequences: Opened the interior of South America to European colonization

There are very large cities that glistened white and besides this land is as good, as fertile, and as normal in appearance as our Spain...

Friar Gaspar de Carvajal's description of the land of the Amazons in his *Relacion* (ca. 1542)

© Anton_Ivanov | Shutterstock.com

The navigation of the Amazon River by Francisco de Orellana ranks among the great epics of exploration, yet it was entirely accidental. The story of the voyage is a rich and fantastical narrative of danger and discovery, the possible significance of which is only recently coming to be appreciated.

The extraordinary success of Francisco Pizarro and his small band of conquistadors in despoiling the Inca Empire and winning a king's ransom in gold served to whet the appetites of gold-crazy adventurers flooding into the New World. Legends proliferated of wealthy civilizations hidden deep within the South American interior — above all of the fantastic El Dorado, the Gilded One. The name El Dorado had originally been applied to a legendary Indian king, who was covered in sticky gum and gold dust as part of a religious ritual that included vast quantities of golden votive offerings. As the conquistadors searched for the source of this legend, El Dorado became synonymous with a city or kingdom of fabulous wealth, greater even than the Inca, driving the lust and ambition of a succession of European adventurers. They launched a series of *entradas*, or expeditions, into the interior, in search of El Dorado, and although the elusive prize seemed always to be just beyond reach there were enough hints and rumors to fan the flames of obsession.

EL DORADO

This was the situation in 1541, when Pizarro, conqueror of the Incas and a powerful figure in the New World, put together an entrada under the leadership of his half-brother Gonzalo, to cross the Andes and search for El Dorado, as well as scouting areas that might be rich in cinnamon and precious metals. Serving as Gonzalo Pizarro's lieutenant was de Orellana, a tough, experienced conquistador who originally hailed from the hardscrabble Spanish region Extremadura and had served under Francisco Pizarro in the 1530s. He may have been related to the Pizarros, an important qualification in the clannish world of competing power blocs seeking to carve out chunks of the New World. Orellana himself had been rewarded for his earlier endeavors (and sacrifices — he had lost an eye during the conquest of the Incas) with vast tracts of land in what is now Ecuador, having been made governor of Guayaquil.

In February 1541, the expedition struck east from Quito. Outfitted at great expense, it consisted of around 300 Spanish conquistadors, 4,000 native porters, 4,000 pigs to serve as a mobile larder, many horses and llamas, and around a thousand vicious war dogs, animals that had proved effective in terrorizing the natives on previous expeditions. Few of the human, and even fewer of the animal, participants were to survive the experience. After months of hacking through difficult and inhospitable terrain the expedition was exhausted, ragged, and running low on provisions, having found nothing of value and only rumors of distant wealth. The party spent Christmas 1541 camped by the upper reaches of what is now the River Coca, and it was agreed that de Orellana should take 50 men and the brigantine—a river-ship built to carry the heavy gear—and scout downstream for provisions; he departed on December 26th. Given the fast-flowing river current and the difficulty of traveling upstream by water or on foot, it was probably unrealistic to expect him to return, and indeed Pizarro and de Orellana would never see each other again.

BREAKING POINT

After just a few days travel de Orellana's party had covered 200 leagues (about 520 miles or 840 kilometers) and reached the confluence of the Coca and Napo rivers, where they found a village of friendly Indians, which was apparently called Aparia—de Orellana was notable for his relatively cordial dealings with Indians, and he made efforts to gather information where he could, but it is not clear how he communicated with the people he met or how reliable his interpretations were.

Here, refreshed and relieved, de Orellana and his men reached a crisis. According to his account it was his men who refused to attempt to return to Pizarro and the main expedition, against de Orellana's own wishes. He made sure to get signatures on a document to this effect but it did not prevent Pizarro from accusing him of faithlessness and betrayal. Undoubtedly, however, it was the sensible choice to make.

At Aparia the breakaway expedition made plans to build a bigger boat, and set about forging nails, which in turn entailed making charcoal. The villagers grew restive with their increasingly demanding guests and the conquistadors decided to push on downriver, soon entering the mainstream of the Amazon. Eventually, they reached another friendly village, which they also, confusingly, named Aparia, and here they built a sturdier brigantine for the challenges ahead. According to the surviving account of the expedition, the *Relacion* of Friar Gaspar de

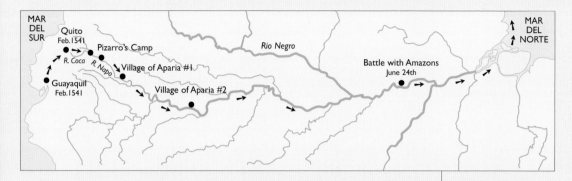

Carvajal, a Dominican friar traveling with de Orellana, evidence of extensive human habitation lay thick along the river banks. He also told of rumors and reports from Indian informants of great kings and their kingdoms all along the river, and he reported eyewitness accounts of several large towns and cities: "we passed through very great settlements and provinces."

Until very recently Carvajal's account has been dismissed as part fantasy, part exaggeration, and part misreporting. The received wisdom was that he must have been exaggerating because the Amazon region is constitutionally incapable of supporting urbanization beyond temporary hunter-gatherer slash-and-burn settlements. This view was influentially advanced by American archaeologist Betty Meggers in her 1971 book *Amazonia: Man and Culture in a Counterfeit Paradise*, which analyzed the nutrient content of the acidic Amazonian soils to "prove" that it could not sustain significant population densities. Yet more recent archaeology, including evidence from satellite and aerial imagery, is suggesting a radical rethink, with traces of extensive habitation and cultivation discovered beneath what was thought to be virgin rainforest—de Carvajal may not have been exaggerating at all. In this case de Orellana's voyage assumes new dimensions of historical interest and importance; on the one hand its recorded testimony offers a unique glimpse at a vanished civilization, but on the other it raises the possibility that the voyage itself was the agent of that civilization's downfall. Later explorers would find little trace of the thriving populations de Carvajal describes or the complex geopolitical patchwork his account (which listed 39 "overlords") implied. What happened to them? It is likely that they were wiped out by a horrific smallpox pandemic, which may have

BOAT TRIP
Orellana's route downriver via the Coca, Napo, and eventually Amazon rivers, finally reaching Marajo Island in the mouth of the great river before turning north and heading up the Atlantic coast.

PESTILENCE IN PARADISE

originated with de Orellana and his band, or perhaps had already been introduced by earlier conquistadors. It is believed that up to 90 percent of Amerindians were killed in the worst pandemic in history.

LAND OF THE AMAZONS

It was now May, and de Orellana and his men were about to enter the most dangerous phase of their journey, borne by the current into the lands controlled by a powerful tribe of female warriors, according to native informants. The Spanish immediately made the link with the Amazons of Greek myth. Venturing ashore for supplies, the conquistadors became embroiled in a vicious fight of over an hour with opponents who "did not lose heart, but it seemed that their courage grew, though many Indians were killed." Their morale was apparently boosted by the arrival of "ten or twelve" Amazons; "we ourselves saw these women and they came and fought before the Indians as captains." De Carvajal described them as "very white and tall," and so skilled with the bow that by the time they had escaped downriver, "our boats looked like porcupines."

They hurried along the lower reaches of the river, meeting hostile Indians at every turn. Near the mouth of the Amazon they encountered Caribs armed with poison arrows, but pushed on, reaching the Atlantic Ocean on August 26th, 1542, exactly eight months after leaving Pizarro's party. Making their way north, they reached a Spanish settlement on September 11th. Back in Spain, de Orellana successfully defended himself against allegations of betrayal and mutiny, and was awarded a grant to conquer and settle the regions he had passed through, now designated as "New Andalusia." In 1546 he returned to the Amazon with four ships but the expedition proved to be a disaster and de Orellana was killed by Indians.

A STAR IS BORN: TYCHO BRAHE AND THE NOVA

1572

Discoverer: Tycho Brahe

Circumstances: A young Danish nobleman noticed a new star in the heavens

Consequences: Ptolemaic cosmology was further discredited; inspired Brahe to follow a career in astronomy

… a new and unusual star, surpassing the other stars in brilliancy, was shining almost directly above my head.

Tycho Brahe

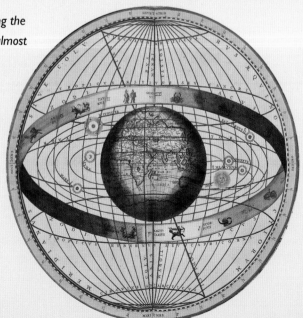

On a cold but clear winter's night in Denmark, a young nobleman emerged from a stuffy laboratory, glanced up at the sky, and saw something impossible. A new star, or nova, blazed in the heavens. Its existence contradicted some of the most fundamental assumptions about the nature of the cosmos, threatening to overturn centuries of academic teaching and religious dogma.

A STAR IS BORN

The nova was visible to most of the people on the planet, and Tycho Brahe (the partially Latinized name of the young Danish nobleman), was by no means the first or only one to notice its extraordinary appearance. Its existence had already been noted by an astronomer named Schuler on November 6th, and for anyone with even a passing acquaintance with the heavens it must have been almost impossible to miss. By the time Tycho saw it, the nova was as brilliant as Jupiter and it grew in brightness until it outshone Venus; for two weeks it was visible in the daytime. Toward the end of November it began to fade, changing color from white to yellow, and fading from orange to a faint red, but lingering on for a year and more until it finally faded from view in March 1574; the nova had been visible to the naked eye for 16 months. Of all those who saw it, however, perhaps none was better equipped to turn an observation into a discovery with implications that could change the prevailing view of the cosmos.

TYCHO BRAHE
Portrait of Tycho Brahe, with implements of astronomy but not, disappointingly, his silver prosthetic nose.

Tycho Brahe's real name was Tyge. In 1546 he was born into an influential Danish family, linked to the council that advised the king. In 1559 he began studying law at the University of Copenhagen, and it was here that his interest in astronomy first flowered. He was struck by an eclipse of August 21st, 1560, and particularly by the fact that it had been predicted. In 1562, now studying at Leipzig, he started to make observations of his own, including a conjunction of Jupiter and Saturn, noting the disparity between the dates predicted for this event by the star tables of Ptolemy and Copernicus. Ptolemy was the ancient astronomer and geographer whose interpretation of Aristotle's cosmology, and particularly its geocentric (Earth-centered) model of the cosmos, had been adopted as dogma by the Church and the scholarly world. The prediction for the conjunction in the Ptolemaic tables, based on a geocentric cosmos, was off by months; even the tables of Copernicus, the Polish priest and astronomer whose heliocentric model had stirred intense controversy, proved incorrect by days. Tycho resolved to do better, aware that the key to improved astronomy would be more accurate observations, which in turn depended on the quality of the instruments used. Improved instruments and unparalleled accuracy of observation were Tycho's great legacy to science, and they first made their mark in his work on the new star that appeared in 1572.

After Leipzig Tycho studied at Wittenberg and Rostock, where in 1566 a duel with a fellow student left him with a part of his nose missing, and in commissioning a gold and silver prosthetic to help cover the disfiguring scar, he conceived an interest in alchemy. Over the next three years he advanced his skill in making and handling astronomical instruments (the telescope had not yet been invented), but also pursued alchemical research, and after the death of his father in 1571 he and his uncle constructed an observatory and laboratory at Herrevad Abbey in Denmark. It was here that he had been laboring on the night of November 11th, 1572, before stepping out and seeing a remarkable sight in the constellation Cassiopeia: "I noticed that a new and unusual star, surpassing the other stars in brilliancy, was shining almost directly above my head." What Tycho was seeing, although he could have no conception of it, was a supernova type 1a: a colossal stellar explosion caused when one of a pair of stars (known as a binary star) collapses into a white dwarf, a massive object that draws ever increasing amounts of hydrogen gas from its neighbor, crushing it until it explodes in a colossal thermonuclear explosion that burns as bright as a billion stars, making it visible across the universe.

Some 10,000 light years away, on Earth, the nova was about to cause a cosmic explosion of another kind. To understand the impact of what Tycho was about to discover, it is necessary to understand the prevailing cosmology of the era. In medieval Europe (and the Islamic world too, to a large extent), Aristotle was the first and last authority on natural philosophy. His physics, which posited the necessary existence of a prime mover or first cause, meshed well with Christian/monotheistic teachings, and his cosmology, as expounded and perfected by Claudius Ptolemy of Alexandria in the 2nd century CE, was received as dogma.

A NOSE FOR TROUBLE

THE HEAVENLY SPHERES

Aristotle had taught that the Earth is the center of the universe, with the heavenly bodies (the Moon, Sun, and planets) and stars rotating around it fixed in vast, nested crystal spheres. The earthly realm of dirty, disordered life, including the atmosphere and all its phenomena, is separate from these crystal spheres, giving a clear distinction between the corrupt and chaotic sublunary realm and the awesome, serene, crystalline perfection of the heavens. A crucial ancillary to this model was that the heavenly spheres were immutable, and to rescue the Ptolemaic cosmology from challenges such as meteorites and comets, it was asserted that they were sublunary atmospheric phenomena. The new star posed just such a challenge.

Tycho made a detailed study of the new star, tracking minutely its position every night. He published the results in a 1573 book, *De Nova et Nullius Aevi Memoria Prius Visa Stella* (On the New and Never Previously Seen Star). Using his unparalleled skill at astronomical observation he had discovered that the new star was farther away than the Moon, and so could not be a mere atmospheric phenomena. Something new had appeared in the supposedly unchanging heavenly sphere, and according to Albert Van Helden, Lynette Autrey Professor of History at Rice University, Tycho's discovery "destroyed the dichotomy between the corrupt and ever changing sublunary world and the perfect and immutable heavens."

Tycho's work was a massive blow to the creaking Ptolemaic cosmology, and by extension the prevailing Aristotelian dogma of natural philosophy. The eventual collapse of the Ptolemaic cosmology would bring crashing down the whole medieval edifice of received wisdom, a necessary precondition for the flourishing of the Scientific Revolution to come. Although the discovery that the nova was superlunary was not the final nail in the coffin of the Ptolemaic cosmology, Tycho would spend the rest of his career hammering in more. The nova was certainly a pivotal moment, not least in convincing Tycho to devote his full career to astronomy rather than alchemy. Thus, the moment at which he stepped out of the laboratory and looked up into the night sky proved deeply symbolic for his future life.

CHALLENGING ARISTOTLE: STEVIN'S FALLING BODIES

1586

CATEGORY

Astronomy

Biology

Chemistry

Exploration

Mathematics

Medicine

Physics

Technology

Discoverer: Simon Stevin

Circumstances: Challenging Aristotle's theories about falling bodies, Stevin and his friend Jan de Groot lug leaden balls to the top of the New Church in Delft

Consequences: Experimental science; laws of motion and gravity

… it will be found that the lighter will not be ten times longer on its way than the heavier, but that they will fall together onto the board so simultaneously that their two sounds seem to be as one.

Simon Stevin,
Waterwicht (1586)

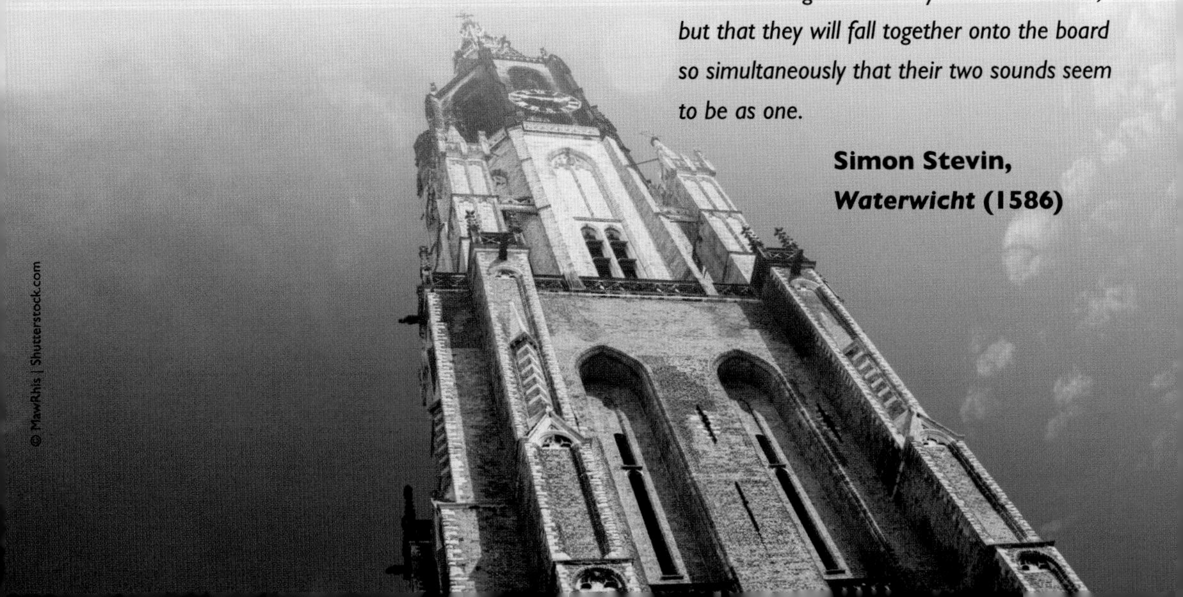

Simon Stevin stands next to a wooden board at the foot of the spire of the New Church, the tallest building in Delft, and signals to his friend Jan de Groot, who is leaning from a window 30 feet (9 meters) above. De Groot simultaneously releases two leaden balls, one ten times the size of the other. They hurtle down from the spire, and the loud thumps that each one makes as it hits the wooden board come so close together that Stevin cannot distinguish one from the other. "So much for Aristotle," thinks Stevin, with grim satisfaction. He has just discovered that 2,000 years of received wisdom are wrong, and performed possibly the first experiment in the history of science to boot.

COMMON SENSE PHYSICS

If you had never heard of gravity and had to come up with a reasonable explanation for why things fall to the ground, you might well arrive at a similar conclusion to the ancient Greek philosopher Aristotle. You might observe that heavy objects seem to land harder when they hit the ground, and you might quite reasonably conclude that the rate at which things fall is proportional to their weight, which is to say, heavier things fall faster. This is common sense physics, at least up to a point, and its apparent logic may help to explain why Aristotle's contentions about falling bodies went unchallenged for 2,000 years. More generally, it was assumed that there was no need to question the theories of Aristotle because he was viewed as the ultimate authority in all of natural philosophy. This unquestioning acceptance of authority was fundamental to the overarching intellectual framework of the Middle Ages: authority was not to be questioned, whether in the religious sphere (the authority of the Church), the temporal sphere (the authority of rulers), or the philosophical sphere — the authority of Aristotle and his derivatives.

© INTERFOTO | Sammlung Rauch | Mary Evans

SIMON STEVIN
Simon Stevin, Dutch polymath and engineer, best known for introducing decimals to European mathematics.

At the end of the Middle Ages, however, in the period known by historians as the Early Modern Era, a new climate of skepticism began to erode these old certainties. In the religious sphere, the Reformation had struck at the authority of the Catholic Church, and this had its echo in the temporal sphere: for instance, in the secession of the northern, mainly Protestant, Netherlands from Catholic Spanish overlords. The Dutch engineer and mathematician Simon Stevin was a key figure in this latter development, helping the rebellious Dutch prince Maurice of Nassau to wage a successful war of independence against Spanish armies. He would also strike a blow against the Aristotelian dogma of natural philosophy with his discovery relating to falling bodies.

DO IT YOURSELF

Aristotle must be counted as one of the most influential people in history, given the length of time during which his ideas were taken as gospel. To medieval scholars he was simply known as "the Philosopher." His assertion that heavy objects fall faster than light ones was taken to be true by the vast majority, educated people included, until the 6th century CE, when the Byzantine scholar John Philoponus was the first to cast doubt on the claim. In a commentary on Aristotle he wrote: "If you let fall from the same height two weights of which one is many times as heavy as the other, you will see that the ratio of times required for the motion does not depend on the ratio of the weights, but that the difference in time is a very small one."

By the 16th century skepticism about Aristotle was becoming more widespread, and more people were starting to test his assertions. In 1575 Girolamo Borro, one of Galileo's teachers at Pisa, tried comparing the fall of wooden and leaden balls, recording that the wooden one fell fastest. The following year Giuseppe Moletti, Galileo's predecessor in the chair of mathematics at the university of Padua, compared the fall from a high tower of balls of different weight and composition, finding that balls released simultaneously reached the ground at the same time. In other words, Stevin was not the first to test Aristotle on falling bodies, nor the first to disprove him. He is, however, regarded as the first to do a proper experimental test, with an attempt to look at the phenomenon of falling bodies quantitatively.

EXPERIENCE AGAINST ARISTOTLE

Simon Stevin was a remarkable polymath; one of the greatest minds of his era, he is peculiarly little known outside Holland, yet deserves to be celebrated alongside near contemporary figures such as Galileo and Descartes. Unlike many more celebrated names, Stevin combined his theoretical genius with practical and mechanical genius; he was a man of action as well as intellect. Of illegitimate birth, he was raised as a Calvinist (a form of Protestantism) and accordingly his family moved to the northern Netherlands to escape Spanish Catholic oppression. Having worked as a book-keeper and clerk he moved to Leiden and joined the University there in 1583, and it was here that he became friends with, and mathematics tutor to, Maurice of Nassau, who would shortly become the leader of the United Provinces (the rebellious northern Netherlands). Stevins advised Maurice on matters of military fortification and engineering, for instance devising a system of sluices to flood the lowlands to hinder enemy armies. Today he is best remembered for his mathematical innovations, such as introducing decimal use and notation to Europe, and for his design for a land yacht, which he built in 1600.

With the help of his friend, burgomaster Jan de Groot, he carried out the experiment that he reported in his 1586 book, *Waterwicht*: "My experience against Aristotle is the following. Let us take… two spheres of lead, one ten times larger and heavier than the other, and drop them together from a height of 30 feet onto a board… Then it will be found that the lighter will not be ten times longer on its way than the heavier, but that they will fall together onto the board so simultaneously that their two sounds seem to be as one." Aristotle's theory predicts that the larger sphere should fall ten times faster than the smaller; Stevin had achieved a rigorous, quantified test of the theory and disproved it.

FROM PISA TO THE MOON

Stevin's experiment was far from the last word on this matter. Three years later Galileo would be trying very similar experiments of his own, as part of the most systematic and sophisticated enquiry yet performed in science. Galileo is famously said to have performed demonstrations similar to Stevin's by dropping balls from the Leaning Tower of Pisa, although this is widely suspected to have been a story made up or embellished by his disciple Vincenzo Viviani. However, it is known that Galileo made multiple attacks on the Aristotelian theory relating to falling bodies. First, he suggested thought experiments that showed up contradictions inherent to the theory; a scenario where falling light and heavy balls are connected by a chain contrasted with one where the two balls are stuck together. In the former case, the Aristotelian version suggests that the lighter ball will slow down the heavier one by holding it back, while in the latter scenario the composite object will be heavier and thus should fall faster than either ball on its own, leading to the patently absurd conclusion that the lighter object both slows down *and* speeds up the fall of the heavier. Second, Galileo conducted a series of experiments substituting the vertical fall of balls dropped from a tower with rolling balls down an inclined plane, slowing the "fall" dramatically and so making it possible to obtain precise measurements of distance covered over time, and thus calculate actual speeds and acceleration. In 1604–1607 this allowed him to formulate his law of falling bodies, which states that the distance traveled by a falling body is directly proportional to the square of the time it takes to fall. But there is also evidence that he genuinely did attempt some version of the famous Pisa experiment, in the precision of his reported observations on the test. When simultaneously releasing a wooden and a metal ball, Galileo claimed that the wooden one moves faster to begin with. This was long taken as proof that Galileo's report was imaginary, but in 1983 Thomas Settle showed that this result can be caused by an experimenter effect—a human experimenter tends to release a lighter ball more quickly, because it is gripped with less force than the heavier ball. Thus, although Galileo thought that he had released both balls simultaneously, in fact he had inadvertently given the wooden ball a head start. The presence of this experimenter effect in his results suggests that he really did perform the experiment.

FABLED EXPERIMENT
Recreation of the scene supposed to have occurred atop the Leaning Tower of Pisa sometime around 1620, with Galileo dropping balls to Earth.

© Hulton Archive/Stringer | Getty Images

Galileo's research program on falling bodies shows that he understood the phenomenon and its significance better than anyone before him. He experimented with multiple materials and even found a way to reduce the influence of friction on the results by using pendulums. Finally, in his 1638 *Discourses and Mathematical Demonstrations Concerning Two New Sciences*, he was able to conclude: "If one could totally remove the resistance of the medium, all substances would fall at equal speeds." He was proven right in 1660 when Robert Boyle used a newly available air pump to watch the fall of a feather and a coin in a vacuum; they fell at the same rate. This same experiment was dramatically recreated on the surface of the Moon in 1971 when astronaut David R. Scott demonstrated live on TV that a hammer and a feather hit the lunar surface at the same time.

CATEGORY

Astronomy

Biology

Chemistry

Exploration

Mathematics

Medicine

Physics

Technology

COURTING CONTROVERSY: GALILEO GLIMPSES THE JOVIAN MOONS
1609

Discoverer: Galileo Galilei

Circumstances: Galileo builds the most powerful telescope yet made

Consequences: Galileo lays the foundation stones for a true science of astronomy and a new conception of physics and the universe; he also incurs the wrath of the Inquisition

… it appears that there are around Jupiter three wandering stars up to this time invisible to one and all.

**Galileo's notebook
(January 11th, 1610)**

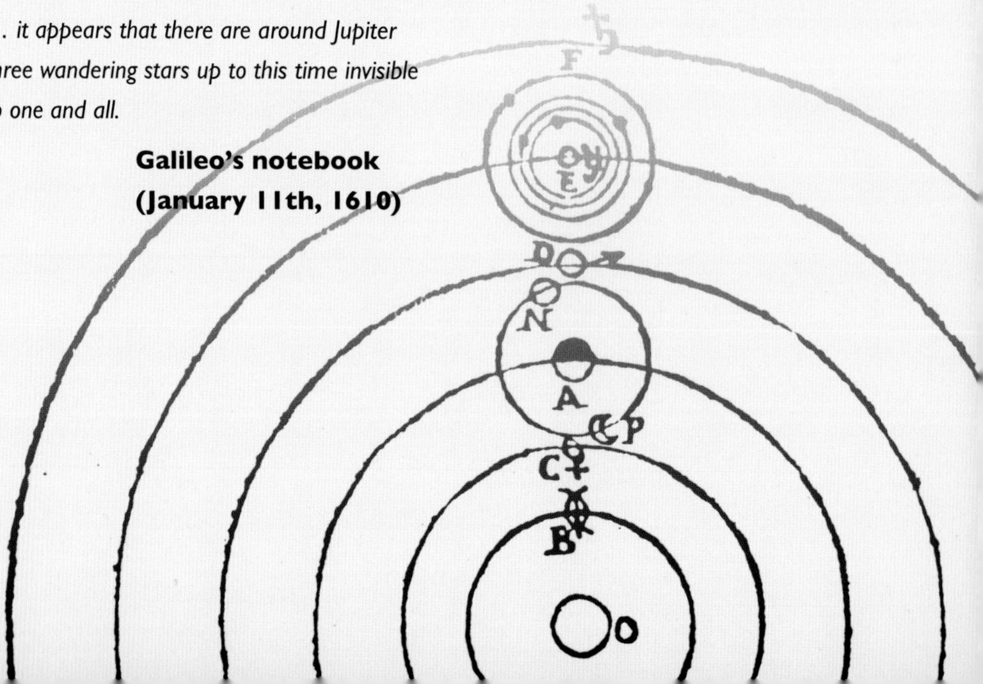

In early 1610 Galileo became the first human to see the moons of Jupiter, a discovery with profound consequences for astronomy, physics, the history of science, and for his own fraught dealings with the Catholic Church and the dreaded Inquisition.

After spending most of December 1609 observing the Moon, Galileo turned his telescope to what, at the time, was the next brightest body in the night sky: Jupiter. The intensely bright but minute object proved difficult for his telescope to handle, but through trial and error he learned to reduce the aperture at the objective. On January 9th he fixed his sights on the planet, noticing three bright little stars in a line on either side of it "of which none was visible without the telescope." Knowing Jupiter's movements across the night sky, he trained his telescope on the planet the next night, expecting to see that it had moved to the west of the little stars; in fact, he observed the opposite: they were still there, strung out in a line to the west of Jupiter. "My perplexity was now transformed into amazement," he recorded in his 1610 book *Sidereus Nuncio* (The Starry Messenger). "I was sure that the apparent changes belonged not to Jupiter but to the observed stars, and I resolved to pursue this investigation with greater care and attention."

Galileo was seeing things never seen before by human eyes, thanks to the marvellous device he had built for himself, which would shortly become known as a telescope. Two years earlier one, or possibly several, Dutch spectacle-makers had invented "a certain device by means of which all things at a very great distance can be seen as if they were nearby, by looking through glasses," according to Hans Lipperhey, one of the men who applied for a patent. The patent was denied, on the basis that the relatively simple design of the device could not be kept secret, and indeed, once word got out, scholars and craftsmen around Europe began to construct their own. By late 1609 Galileo had built several, achieving at least eight times magnification, and he quickly mastered the subtleties of getting such a delicate instrument to perform. For a period of nearly a year Galileo had access to an instrument more powerful than anyone in the world, and he made good use of his technological head start.

He began by examining the Moon. In the Aristotelian cosmology the Moon was part of the perfect heavenly realm, made from perfect

FROM PERPLEXITY TO AMAZEMENT

THE MOON'S NOT A BALLOON

and incorruptible ethereal substance, and must of necessity be a perfect, smooth sphere. The darker patches on the face of the Moon had traditionally been explained as variations in density of the lunar substance. Galileo observed through his telescope new shapes and patches not visible to the unaided eye, and watched as they changed shape like shadows cast by ridges and mountains, concluding that this must be exactly what they are. In other words, the Moon is not perfectly spherical, and at least one body in the heavenly realm is not much different from the Earth. Later observations of sunspots suggested similar imperfections on the Sun.

CONTROVERSIAL INSTRUMENT
Galileo demonstrating the power of his telescope to the Doge of Venice; it would soon become the instrument of much contention.

When he turned his telescope to Jupiter and saw the three "stars wandering about Jupiter" (later spotting a fourth), it took him just four days to conclude that these bodies orbited the planet "as do Venus and Mercury about the sun." Later that year his discovery of the Jovian moons—now known as the Galilean moons, although Galileo himself called them the Medicean moons after his patrons the Medicis—was announced through his book, *The Starry Messenger*, which caused a sensation throughout Europe: "our own eyes show us four stars which wander around Jupiter as does the moon around the earth, while all together trace out a grand revolution about the sun in the space of 12 years…"

These astronomical discoveries proved that there were many more things in the heavens than dreamed of by Aristotle or Ptolemy, the authors of the established cosmological dogma, striking at their authority and thus the basis of their legitimacy. The discovery of the Jovian moon system also had much more profound consequences, and to understand why it is necessary to look at the integrated nature of Aristotle's physics and cosmology.

Aristotle believed that motion derived from the nature of objects, so that earthly objects fall down toward the center of the Earth because of their "earthy" nature. Meanwhile, the heavenly bodies move in perfect circles because they are composed of some empyrean substance. The earthly and heavenly realms are utterly distinct; on Earth, things move in straight lines, while in the heavens they have circular motion—in other words, earthly and heavenly physics are different. These premises lead on to the conclusion that the Earth must be the immobile center of the universe, about which all the other bodies revolve. The Earth must be center, because earthy substances move toward it (i.e. they fall when dropped), and if there were two centers (as suggested, for instance, by the heliocentric cosmology, in which the Moon goes round the Earth but the Earth goes round the Sun), then earthy elements would not "know" in which direction to fall and this would affect their observed motion. "From these considerations," wrote Aristotle in his influential treatise of 350 BCE, *On the Heavens*, "it is clear that the earth does not move and does not lie elsewhere than at the center."

Jupiter and its moons shattered this entire system at a stroke. Here was proof that there existed other "centers of circular motion" (i.e. of gravity) in the cosmos. The same physics appeared to apply in the heavens as on Earth, and the case for the Copernican, heliocentric cosmos was immeasurably strengthened. If Jupiter and its moons could move around the Sun, why not the Earth and its moon? In December 1610, Galileo would make yet another discovery of cosmological significance, observing that Venus had phases just like the Moon. This was the first definitive proof that Venus went round the Sun, not the Earth, and the greatest support that the Copernican system had yet received.

EARLY OBSERVATIONS
Draft for a letter to the Doge of Venice relating to Galileo's telescope; Galileo used the bottom of the sheet for sketches of his first observations of the Jovian moons in early January 1610.

Galileo operated in the rough and tumble world of academic dispute and carried on a number of vicious disputes with other scholars. One enemy with whom he dared not tangle, however, was the Catholic Church. And yet Galileo enjoyed the patronage of Cardinal Maffeo Barberini, who would later become Pope Urban VII, and felt secure enough to complain about those, including learned clergy, who would not even look at his new discoveries for themselves. "My dear Kepler," he wrote in 1610, "what would you say of the learned here, who, replete with the pertinacity of the asp, have steadfastly refused to cast a glance through the telescope? What shall we make of this? Shall we laugh, or shall we cry?"

The theories he was expounding ran contrary to Church doctrine, particularly when he wrote a letter suggesting that scriptural assertions

EXPECT THE INQUISITION

that the Earth did not move should not be taken literally; accusations were leveled against him. In 1615 Dominican friar Father Lorini wrote to the Inquisition to complain: "the letter contains many propositions which appear to be suspicious or presumptuous, as when it asserts that the language of Holy Scripture does not mean what it seems to mean… that the Holy Scriptures should not be mixed up with anything except matters of religion…" On February 23rd, 1616, the Inquisition declared Galileo's heliocentric propositions "foolish and absurd" and "formally heretical." He was admonished "to abandon the said opinion… that the Sun is the center of the world and immovable and that the Earth moves," and told that he could only teach Copernicanism as a "hypothesis," not as fact.

In 1623 Barberini was elected Pope and gave Galileo permission to publish a new book on his theories and discoveries, the *Dialogue Concerning the Two Chief World Systems*, but the scientist overstepped the mark in his propounding of Copernicanism and the Pope was angered. The Inquisition accused Galileo of breaking the injunction of 1616 and in 1633, under threat of torture, he was forced to abjure his beliefs and was sentenced to spend the rest of his life under house arrest. "I curse the time devoted to these studies in which I strove and hoped to move away somewhat from the beaten path," he lamented bitterly in 1633. "I repent having given the world a portion of my writings; I feel inclined to consign what is left to the flames and thus placate at last the inextinguishable hatred of my enemies."

SUPER FAST MATH: NAPIER'S LOGARITHMS

1614

CATEGORY

Astronomy

Biology

Chemistry

Exploration

Mathematics

Medicine

Physics

Technology

Discoverer: John Napier

Circumstances: Advances in astronomy and mathematics increasingly demanded exhaustive and tedious calculations; a means was needed to make calculating faster and easier

Consequences: Radically accelerated the development of mathematics and astronomy

Seeing there is nothing… that doth more molest and hinder calculators, than [calculations] of great numbers, which besides the tedious expense of time are for the most part subject to many slippery errors, I began therefore to consider in my mind by what certain and ready art I might remove those hindrances.

John Napier, *Mirifici logarithmorum canonis descriptio* (1614)
(from the preface of the first English translation in 1616)

John Napier's discovery of logarithms—a word he coined from Greek roots meaning "reckoning numbers"—cost him at least 20 years of mathematical labor. For most people today, logarithms are not encountered outside the classroom, so it is hard to imagine the impact they had at the time; the most obvious analogy is with the modern electronic calculator, an instrument so synonymous with its function that the "electronic" is rarely used.

SECRET WEAPON FOR THE WAR WITH MARS

The adoption of the Hindu-Arabic numerals in the Middle Ages had opened the way for radical change in European mathematics, which had failed to advance since the Classical period. The numerals made possible calculations that had been impractical or even impossible with Roman numerals, but these calculations in turn began to grow as the mathematicians became ever more ambitious. A typical example is the struggle endured by Kepler as he attempted to calculate the orbit of the planet Mars using data from the astronomer Tycho Brahe, a labor he later described as "my war with Mars," which resulted in him filling over a thousand sheets with arithmetical calculations.

Such computation was, in the words of John Napier himself, "hard and tedious" and inevitably "subject to many slippery errors." Accordingly, mathematicians searched for ways to "save computation," primarily by turning multiplication and division into much more straightforward additions and subtractions. In 1510 Johannes Werner recognized that trigonometric identities (such as sine and cosine) could be used in this fashion, in a technique that came to be called prosthaphaeresis, from the Greek *prosthesis* (addition) and *aphaeresis* (subtraction), but it was not until 1588 that an application of this nature first appeared in print.

Not long after this the problem of saving computation began to exercise the mind of John Napier, laird of Merchiston, a Scottish landowner and theologian who made a hobby of mathematics. In 1614 he was to publish *Mirifici logarithmorum canonis descriptio* (A Description of the Wonderful Table of Logarithms), but by his own account he had been working on the problem for at least 20 years before this. His original inspiration may have been a report from his friend the royal physician Dr. John Craig, who had accompanied James VI of Scotland to a dinner at Tycho Brahe's castle and there heard how the astronomer and his assistants used prosthaphaeresis to simplify their calculations.

Most of Napier's research on the topic had been carried out while he was living at his estate at Gartness, where his house adjoined a mill with a noisy race or "cascade." According to the *Statistical Account* (a historical and demographic record of Scotland compiled in the 18th and 19th centuries), "the noise of the cascade, being constant, never gave him uneasiness, but that the clack of the mill, which was only occasional, greatly disturbed his thoughts. He was therefore, when in deep study, sometimes under the necessity of desiring the miller to stop the mill that the train of his ideas might not be interrupted."

The fruit of his labors, which included calculating ten million entries, from which he compiled the tables that lay at the heart of his new invention, was his 1614 *Descriptio*. In the preface to the English translation of the Latin original, published in 1616, Napier explains how his tables "together with the hard and tedious multiplications, divisions, and extractions of roots, doth also cast away from the work itself even the very numbers themselves that are to be multiplied, divided, and resolved into roots, and putteth other numbers in their place which perform as much as they can do, only by addition and subtraction, division by two or division by three."

A logarithm is the reverse of an exponential. A simple example of an exponential is 2^3 = 8, where the 2 is known as the base and the 3 is the exponent or power, so that 2^3 = 8 in words becomes "base 2 raised to the exponent 3 equals 8." The opposite—or inverse—of this is "base 2 logarithm of 8 equals 3," written in mathematical notation as $\log2(8)$ = 3. The term in parantheses (in this case 8) is known as the "argument." A logarithm can be thought of as the answer to the question "To what power must the base be raised in order to give the argument?" So the $\log2(16)$ translates as "To what power must 2 be raised to give 16?," and the answer is 4—i.e. 2^4 = 16.

Logarithms are useful for simplifying calculations because they relate geometric series to arithmetical ones. A geometric series is one that increases exponentially—for example, the series 2, 4, 8, 16, 32, 64, 128, 256, 512, 1,024. An arithmetic series is one that increases through simple addition, such as 1, 2, 3, 4, 5, etc. These two series are linked

© Print Collector | Getty Images

AN INVENTIVE MIND
John Napier, the Laird of Merchiston, whose inventive mind concocted a variety of fiendish devices of war, in addition to his famous logarithms.

UNDERSTANDING LOGARITHMS

because the latter is composed of the base two logarithms of the former; putting them together gives a simple table:

1	2	3	4	5	6	7	8	9	10
2	4	8	16	32	64	128	256	512	1,024

The bottom row shows the powers of 2 and the top row shows the exponents to which 2 must be raised to give these powers. This table can be used to make a complicated multiplication into a simple addition, e.g. 8 × 64. Use the table to look up the exponents corresponding to each term, to get 3 and 6; add these together to give 9; and look up the corresponding power, which is 512. This gives the answer 8 × 64 = 512, but requires only addition. This, in microcosm, is the system created by John Napier, although his logarithms differed from the modern understanding. This table itself was used as an explanatory aid by the Swiss Joost Burgi, who independently invented logarithms but did not publish his tables until 1620.

With the tables of logarithms published by Burgi and by Napier and his colleague Henry Briggs (see below), a host of difficult calculations could be dramatically simplified. For multiplication, you look up the log that corresponds to your starting numbers, add them together, and then look up the antilogarithm of the sum to find the product of their original multiplication. For division, one log is subtracted from the other, and then the antilog of the result is looked up. Logarithms also simplify finding squares and cubes, and square and cube roots. To find the square of a number, look up its log, multiply it by two and then look up the antilog of the answer; to find the cube root of a number, divide its log by three and look up the antilog.

ENGINE OF WIT

One of Napier's biggest fans was the English mathematician Henry Briggs, who wrote to a friend in March 1614 of how Napier's new book "hath set my head and hands a work with his new and admirable logarithms. I hope to see him this summer, if it please God, for I never saw a book which pleased me better or made me more wonder." Briggs did indeed see Napier, making the long and arduous 400-mile (645-kilometer), four-day journey from London to Edinburgh. According to an account of the meeting told by John Marr to William Lilly, when Briggs was shown into Napier's chamber, "almost one quarter of an hour was spent, each beholding other with admiration, before one word was spoken. At last Mr Briggs began, 'My Lord, I have undertaken this long journey purposely to see your person, and to know by what engine of wit or ingenuity you came first to think of this most excellent help unto astronomy, viz. the Logarithms.'"

Alas, the Laird's answer does not survive, but there can be little doubting the strength of his creativity. Napier also invented a host of imaginative, albeit potentially far-fetched, "devices for the ruin and overthrow of man," including "devices of sailing under the water… a closed and fortified carriage to bring arquebusiers into the midst of an enemy… [and] a kind of shot for artillery… calculated to clear a field of four miles' circumference of all living things above a foot in height." By it, he said, the inventor could destroy 30,000 Turks, without the hazard of a single Christian. Napier also popularized the use of the decimal point, explaining that, "In numbers distinguished thus by a period in their midst, whatever is written after the period is a fraction, the denominator of which is unity with as many ciphers after it as there are figures after the period." But his greatest invention and gift to posterity was the logarithm, which the great 18th–19th-century French mathematician and astronomer Pierre-Simon Laplace famously said had "by shortening the labors, doubled the life of the astronomer."

CATEGORY

Astronomy

Biology

Chemistry

Exploration

Mathematics

Medicine

Physics

Technology

STORMING THE DARKNESS OF THE MIND: THE LAWS OF PLANETARY MOTION

1618

Discoverer: Johannes Kepler

Circumstances: Selected by Tycho Brahe to analyze his mountains of data, Kepler seeks the simple mathematical laws that will explain the motions of the planets

Consequences: Lays groundwork for Newtonian physics and the discovery of gravity

If you want the exact moment in time, [the Third Law] was conceived mentally on March 8th in this year one thousand six hundred and eighteen, but submitted to calculation in an unlucky way, and therefore rejected as false, and finally returning on the 15th of May and adopting a new line of attack, stormed the darkness of my mind.

Johannes Kepler, *Harmonice mundi* (1619), Book 5, Chapter 3

The transformation of natural philosophy into science was achieved by combining observational and experimental data with the rigor of mathematics. Through the discovery of quantifiable, mathematically based laws of nature, science gained its remarkable power to explain and refashion the natural world. One of the first such discoveries was Kepler's third law of planetary motions, which would open the way for the discovery of the law of universal gravitation 50 years later.

In his posthumously published 1634 book *Somnium*, the first science-fiction book of the modern era, the German mathematician and astronomer Johannes Kepler told an extraordinary tale of a boy born to a hot-tempered sorceress, who is sent to spend five years in a castle on an island in the far north, apprenticed to a star-gazing magus. Remarkably, this tall tale is largely autobiographical; Kepler's mother was indeed a difficult woman accused of witchcraft (he personally conducted her defense, securing her release in 1620 after five years of extreme jeopardy and the threat of torture). The magus was none other than Tycho Brahe (see page 49), and although Kepler did not work with him at Uraninborg, his castle of wonders in the far north, he did spend a year working as his assistant in Prague—a year that was to change the history of science.

Born in southwest Germany in 1571, Kepler studied at Tübingen University, where he was introduced to Copernicanism, a model he ardently embraced for religious/metaphysical reasons, but also because, as many astronomers were finding, compared to the geocentric Ptolemaic model it gave a much better fit with observations—in other words, with reality. In 1597 he published *The Cosmographic Mystery*, in which he painstakingly attempted to prove that the orbits of the planets around the Sun fit within a series of nested regular solids (circle, square, triangle, etc.). Although this attempt would prove futile, his ingenuity caught the eye of Tycho Brahe, at this time serving as the imperial mathematician to Rudolph II in Prague. Tycho needed someone to analyze his observations of the planets and calculate new orbits; Kepler moved to Prague in 1600 and set to work. Tycho died the following year and Kepler took over his post. There followed many long years of arduous calculations, particularly

THE SORCERER'S APPRENTICE

JOHANNES KEPLER
As well as his famous laws of planetary motion, Kepler made significant strides along the path to calculus, the mathematics of change and curves.

concerned with his attempts to work out the orbit of Mars, a process that Kepler later described as his "war with Mars."

For Kepler, the unprecedented accuracy of Tycho's observations was a gift from God that would enable him to disprove the Ptolemaic system, as evidenced by the significant error in its description of the orbit of Mars, which was wrong by eight minutes. Kepler's privilege of observation over authority marks him out as a scientist in the modern sense, for all his metaphysics. "Because they could not be ignored, these eight minutes, all alone, have opened up the road to reforming the whole of Astronomy, and they have become the material for a large part of this work," he wrote in his 1609 book *Astronomia Nova* (New Astronomy), the fruit of his labors with Tycho's data.

THE MACHINERY OF THE HEAVENS

In the book he explained his deduction that the orbits of the planets were not circles, as had always been assumed by both the Ptolemaic and Copernican hypotheses, but ellipses. He also set out his first two laws of planetary motion. First, that a planet orbiting the Sun follows an elliptical path, with the Sun at one of the foci of the ellipse (the foci are the two points inside the ellipse that define its shape). Second, that as the planet moves around its orbit a line from the focus to the planet sweeps out equal areas in equal times, which is a geometer's way of saying that the speed at which the planet orbits depends on its distance from the Sun (it accelerates as it gets closer).

In setting out these laws Kepler was attempting to discover the secret architecture of the universe and profoundly change the conception of cosmology. In a famous passage from a letter he wrote in 1605, he explains: "My aim is to say that the machinery of the heavens is not like a divine animal but like a clock… and that in it almost all the variety of motions is from one […] force acting on bodies, as in the clock all motions are from a very simple weight." Kepler came tantalizingly close to identifying gravity as this force, but eventually he erroneously settled on magnetism. His discovery of the Third Law, however, would open the door to gravity.

© Print Collector | Getty Images

THE ORBITS OF THE PLANETS

Kepler's diagram depicting his ultimately unsuccessful attempt to show that the orbits of the planets described a series of nested Platonic solids.

For the next 12 years Kepler searched for a deeper pattern in the orbital data he had inherited, striving to break the code of creation and read the deepest secrets of nature. The Pythagorean insistence on the cosmic significance of musical harmonies was a particular inspiration, and he looked for simple ratios in the data. But it was not until after he had read about Napier's discovery of logarithms in 1616 that he finally made the breakthrough. Kepler's old teacher from Tübingen, Michael Maestlin, had teased him about his enthusiasm for logarithms, claiming that "it is not seemly for a professor of mathematics to be childishly pleased about any shortening of the calculations," but it is likely that Kepler's profound understanding of logarithms underlay his breakthrough with the Third Law.

On March 8th, 1618, he later wrote, a new idea "appeared in my head," and he saw that "the proportion between the periodic times [time taken to orbit the Sun] of any two planets is precisely one and a half times the proportion of the mean distances [the average distances of the planets from the Sun]." In graphical terms the Third Law can be shown using a log graph (where the axes are logarithmic), which reveals the simple relationship as a straight line. It is highly likely that Napier's book provided the key. After a false start caused by an error in calculation, Kepler formulated the Third Law on May 18th, 1618, writing: "Now, because 18 months ago the first dawn, three months ago the broad daylight, but a very few days ago the full Sun of a most highly remarkable spectacle has risen, nothing holds me back." Eighteen months before May 1618 would have been just the time when he was reading Napier's *Description of the Wonderful Table of Logarithms*.

A MOST HIGHLY REMARKABLE SPECTACLE

His new discovery fit so well with his labor of 17 years on the observations of Brahe, that at first he believed he was dreaming. But the simple mathematical rigor of the Third Law could not be denied. Where Kepler's first two laws had excited widespread antagonism in the astronomical community, his third was accepted without demur. But its true significance lay unrealized for several decades, until it inspired Isaac Newton.

Kepler made many other discoveries in his life. He was the first to describe accurately the workings of the eye, showing that the lens in a human eye casts an inverted image onto the retina. He worked out the optical principles behind the astronomical telescope and designed an improvement on Galileo's version. His own telescopic observations proved to be an important support to Galileo in the debate over the

Jovian moons (see page 58). On the occasion of his second wedding he was moved to calculate the volume of wine contained within barrels, which in turn led him to create an early form of calculus. The astronomical tables based on the Copernican system he compiled while working as the imperial mathematician, known as the Rudolphine Tables, were so accurate that they constituted the single most powerful argument for the superiority of the heliocentric model, since people all over Europe could see that this model worked the best. Unfortunately, the political-religious disputes convulsing central Europe at this time, which saw Germany plunged into the nightmarish Thirty Years War, led to Kepler losing academic positions and patronage, and he ended his years in financial difficulties, dying in 1630 while en route to Prague to attempt to recover unpaid salary from his time there. He composed his own epitaph: "I used to measure the Heavens, now I measure the shadows of Earth. The mind belonged to Heaven, the body's shadow lies here."

ABHORRENT TO NATURE: THE EXISTENCE OF THE VACUUM

1644

CATEGORY

Astronomy

Biology

Chemistry

Exploration

Mathematics

Medicine

Physics

Technology

Discoverer: Evangelista Torricelli

Circumstances: Galileo begins to explore the nature of the vacuum, but dies soon after; his brilliant new secretary is left to pursue the investigation

Consequences: Settles a 2,000-year-old dispute about the existence of a vacuum

... certain philosophical experiments are in progress... relating to vacuum, designed not just to make a vacuum but to make an instrument which will exhibit changes in the atmosphere...
Evangelista Torricelli (1644)

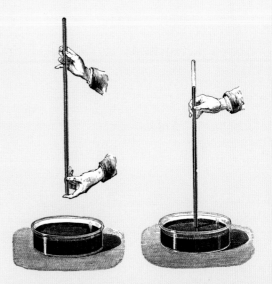

Torricelli's discovery that the vacuum definitely existed, made by way of the invention of the barometer, an instrument for measuring atmospheric pressure, is one of the classic examples of the power of a single elegant experiment.

DOES NATURE ABHOR A VACUUM?

One of the most important influences on the evolution of natural philosophy into science in the early modern period was input from the practical side of life. While medieval philosophy had been the province of ivory towers, where scholars proverbially debated how many angels could dance on the head of a pin, early modern natural philosophy would increasingly concern itself with the real world experience of engineers, miners, and artisans, and seek useful applications of its findings toward the improvement of these fields.

Thus, in 1641, Galileo took on a problem raised by Tuscan engineers who were attempting to devise effective pumps for raising water from mines or wells. The pump makers had run up against some sort of natural limit—no matter how hard they tried, they could not get their pumps to lift water above a height of about 35 feet (just over 10 meters). At this point the water would separate from the pump plunger and rise no higher, leading to the intriguing question: what was the nature of the space between the top of the water and the bottom of the plunger?

No stranger to controversy, Galileo was now straying into yet another contentious field: the question of the vacuum. Aristotle and Plato had agreed the very concept of the vacuum to be self-contradictory: something could not be occupied by nothing. As Aristotelian philosophy hardened into medieval Church dogma, this reasoning came to be summed up in the phrase "nature abhors a vacuum," which first appeared in English in 1550 when Thomas Cranmer, the Archbishop of Canterbury, wrote that "Naturall reason abhorreth vacuum, that is to say, that there should be any emptye place, wherein no substance shoulde be."

Galileo's answer to the pump makers' riddle was a novel take on this principle. Despite being well acquainted with the fact that air has weight, he missed the true explanation for the drawing limit of a pump, suggesting that while the drawing power of the pump was due to nature's horror of the vacuum forcing water up the pipe to prevent the formation of such a vacuum, this horror had a limit, extending only to about 35 feet. His new secretary, a bright young man by the name of Evangelista Torricelli, had other ideas, but Galileo died in January 1642, less than four months after his arrival.

By force of intellect Torricelli had risen from lowly circumstances to become an accomplished mathematician. Among his achievements, he discovered the mathematical figure now known as Torricelli's horn, which he described as "the acute hyperbolic solid." The horn is remarkable because it is infinitely long, but has a finite volume, a finding Torricelli himself described as "incredible," and which was important in the development of calculus as a demonstration of how infinite series can be summed to give finite results.

An avid follower of Galileo, he wrote on the mathematics of motion, which led in turn to his position as the great man's secretary. On Galileo's death, Torricelli succeeded him as mathematician to the Grand-Duke of Tuscany, and professor to the Florentine Academy, a position he held until his tragically early death in 1647.

Continuing researches into the vacuum problem, Torricelli built a startling facsimile of the problematic pump tube, setting up a giant water column in his own home. He must immediately have apprehended that the level of water in the 35-foot-tall tube rose and fell with changes in the weather, and he alarmed the local community by fixing a dummy to the top of the column, so that it too rose and fell with changes in atmospheric pressure.

What Torricelli had created was the world's first barometer (a word from Greek roots meaning "pressure measurer," although it was not called this at the time). Already he had considered that an easier demonstration of the principles involved might be achieved by using a medium heavier than water, namely mercury, which is about 13½ times more dense than water. Accordingly, Torricelli theorized, where the height limit for water is ca. 35 feet, the height limit for mercury should be 13.5 times lower, or around 30 inches (76 centimeters); or to put it another way, the weight of a 30-inch column of mercury is equivalent to that of a 35-foot (10.6-meter) column of water. Such a demonstration would require only a tube that could sit on a table, rather than one that projected above the roof of his house.

EXHIBITING ATMOSPHERIC CHANGES

BRILLIANT MATHEMATICIAN
Torricelli, a brilliant mathematician and scientist whose early death perhaps prevented him from becoming as celebrated as his mentor, Galileo.

BIG BAROMETER
A slightly out-of-scale depiction of Torricelli's barometer experiment, with various long tubes upended into a large tub, presumably full of mercury.

In a letter to his friend Michelangelo Ricci, dated June 11th, 1644, Torricelli described what he and his assistants had done: "We have made many glass vessels… with tubes two cubits long. These were filled with mercury, the open end was closed with the finger, and the tubes were then inverted in a vessel where there was mercury… We saw that an empty space was formed and that nothing happened in the vessel where this space was formed…"

This simple experiment proved that there could be nothing in the space above the mercury — it must be "an emptye place, wherein no substance shoulde be"; in other words, a vacuum. Simply by upending a tube, Torricelli had created a vacuum, supposedly abhorred by nature yet remarkably easily created. "Many have argued that a vacuum does not exist," he commented, "others claim it exists only with difficulty in spite of the repugnance of nature; I know of no one who claims it easily exists without any resistance from nature." Yet this is exactly what he had just demonstrated.

Torricelli went on to explain why the mercury rose to a certain height in the tube: "I claim that the force which keeps the mercury from falling is external and that the force comes from outside the tube. On the surface of the mercury which is in the bowl rests the weight of a column of 50 miles of air. Is it a surprise that… the mercury… should enter [the tube] and should rise in a column high enough to make equilibrium with the weight of the external air which forces it up?" As the precise weight of this 50-mile (80-kilometer) column of air varied with the weather, so the height of the mercury that it could support would vary, making Torricelli's apparatus "an instrument which will exhibit changes in the atmosphere, which is sometimes heavier and denser and at other times lighter and thinner."

THE ENDURING BAROMETER

Torricelli confined announcement of his findings to his letter to Ricci, an old school friend; he was keenly aware of the dangers posed by discoveries that contradicted Church doctrine, having witnessed at first hand Galileo's treatment. Indeed, he had pointedly eschewed the controversial field of astronomy in his studies precisely in order to avoid trouble. Yet news of the discovery spread; in October 1644 Torricelli was visited by the French mathematician and networker Marin Mersenne, who took copies of the Ricci letter back to France, where its significance was immediately grasped by the French natural philosopher Blaise Pascal. Pascal repeated the experiment for himself, and proved his contention that atmospheric pressure decreases with altitude by sending a barometer up a mountain; at the peak, the level of mercury was significantly lower, proving that the weight of air now above the barometer was much less. These findings in turn led Pascal to estimate the total mass of the Earth's atmosphere, arriving at a figure close to modern values. Torricelli himself demonstrated a profound understanding of his discovery of atmospheric pressure, realizing that it explained the nature of the wind. "Winds are produced," he would later tell students in a lecture, "by differences of air temperature, and hence density, between two regions of the earth."

Mercury barometers as devised by Torricelli went on to become standard instruments for scientific, domestic, commercial, and maritime purposes. For over 300 years essentially the same design would be used as the reference standard, which may be a record for scientific instruments. Not until 1977 did the U.S.National Weather Service switch to electronic barometers that use piezoelectric crystals (where the current generated by the crystal depends on the pressure exerted on it).

CATEGORY

Astronomy

Biology

Chemistry

Exploration

Mathematics

Medicine

Physics

Technology

AN ODD STRAYING OF THE LIGHT: GRIMALDI PERCEIVES DIFFRACTION
1660

Discoverer: Father Francesco Grimaldi

Circumstances: Grimaldi makes a strange discovery while investigating the nature of light

Consequences: Suggests that light is wave-like in nature; leads to invention of diffraction grating; inspires Newton; lays the basis for major elements of optical science

... an odd straying of light caused in its passage near the edge of a Razor, Knife, or other opaque body in a dark room; the rays which pass very near the edge being thereby made to stray at all angles into the shadow of the knife.

Isaac Newton describing Grimaldi's discovery in 1675

Is light made up of tiny bits of stuff, or is it like a fluid, which moves in waves? Bizarrely it is both, but the difficulty of reconciling this apparently self-contradictory state of affairs has led to one of the longest-running and most bitter battles in the history of science. The first blow in this battle was struck in a darkened room in Bologna sometime around 1660, where a Jesuit priest made a startling discovery.

Francesco Grimaldi was born in Bologna in northern Italy in 1618 to an affluent family of silk merchants and chemists. He entered the Jesuit order in 1632. The practice of the Jesuits was to put bright initiates through an extensive period of education and academic work before they took their orders and became priests. Grimaldi would not become a priest until 1651, by which time he had won a doctorate, taught in almost every branch of academia, and spent 15 years as assistant to the Jesuit scientist Giovanni Battista Riccioli, contributing to the latter's work on motion, ballistics, and astronomy. This period saw the emergence of both Grimaldi's experimental ingenuity—Riccioli, in his 1651 work *Almagestum novum*, credits him with originating no less than 40 original experiments—and his great facility with accurate measurement.

A RELIABLE MEASURER

In the last ten years of his life Grimaldi devoted much time to exploring the nature of light, summing up his work in the posthumously published 1665 work *Physico-Mathesis de Lumine*. Though much of the book is devoted to an esoteric discussion of whether light is a substance or a quality, it also describes his most significant experiment. The experiment was devised to test the commonly held assumption that light propagates in rectilinear fashion, which was the technical phrase for "moving in a straight line." If light is indeed made up of tiny particles—a position known as the corpuscular theory of light, because it assumes that light comprises tiny bodies or corpuscles—then a ray of light would be expected to consist of streams of particles traveling in straight lines. If this is the case, what should happen when an object is placed in the path of a ray of light, casting a shadow on a screen? If light does go in straight lines, the shadow should have a hard edge, which should fall along a line drawn straight from the object to the screen. This is exactly what Grimaldi set out to examine; like many experiments of the early modern era, this one is remarkable for its simplicity—anyone could have performed it at almost any time in human history.

AN ODD STRAYING

Grimaldi made a tiny hole in the shutter of a darkened room, admitting light that fell onto a screen. Into the cone of light thus admitted, he placed a straight-edged metal bar and carefully observed the shadow it cast. The expected result can be shown with a simple diagram (see opposite), where the hole in the shutter is the gap AB, which throws a cone of light onto the screen CD. If light travels exclusively in straight lines, the edges of the shadow cast by the bar FE should be I and L, but Grimaldi saw something unexpected. The actual shadow cast by the bar ran from M to N, and he saw that "the limit of the shadow remains in some way ill-defined, due to a certain penumbra, characterized by a perceptible graduation... in the regions between total shadow and full light." In the diagram, only the zone GH was in full darkness, with light somehow bleeding into the zones IG and HL. No light can be traveling directly in a straight line from the hole in the shutter to these zones, so the only possible explanation is that somehow light is going around a corner. Isaac Newton later described it as "an odd straying of the light."

What was more, Grimaldi noticed that at the borders of the shadow light appears in "fringes" or bands with brighter light in the center and color at the edges, and that these fringes only appear if the hole in the shutter is small enough. In further experiments Grimaldi even proved the amazing result, "That a body actually enlightened may become obscure by adding new light to that which it has already received." In other words, it is possible to cast a shadow on a previously illuminated object by *adding* light!

DIFFRACTION AND INTERFERENCE

Grimaldi interpreted his initial findings to mean that light must be behaving like a fluid, which moves in waves. Just as waves in the sea hitting an object break and bend around the object, so the light waves must be breaking and bending around the metal bar, which explains why some light "strays" into the zone of shadow. To describe this phenomenon Grimaldi coined the term "diffraction," from the Latin *diffringere*, to "break into pieces." Diffraction is where waves bend around small obstacles, or spread out from small openings (the opening must be near the wavelength of the wave in question, which is why very small slits or holes work best for light, which has very small wavelengths). Diffraction can be seen at work in many aspects of everyday life; for instance, if you look at the surface of a CD or

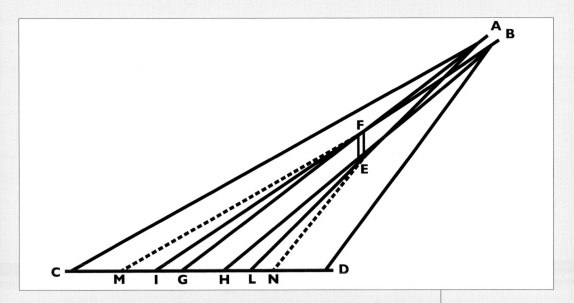

DVD, you will see rainbow-colored patterns. These are caused by light diffracting as it reflects off the surface of the disc and passes through the incredibly thin grooves that make up the tracks.

Although Grimaldi's book and the discoveries described in it were not widely heralded, they did catch the attention of the French Jesuit Honoré Fabri, who in turn was read by Newton. Thus Grimaldi's work helped inspire the young Englishman to undertake his own ground-breaking investigations into optics (see page 82). However, the proper explanation of Grimaldi's findings did not emerge for some time. He had, for instance, discovered the phenomenon now known as interference with his finding that it is possible to cast a deeper shadow by adding light. It is now known that these phenomena conclusively demonstrate that light does indeed have a wave-like nature, and that they result from one wave of light interfering with another. Where the two waves are in synchrony, their peaks add together to produce higher peaks (i.e. more brightness—this is known as constructive interference), and where a peak coincides with a trough, the two cancel each other out to produce darkness (destructive interference). This is why adding light can lead to darkness.

DIFFRACTION
A diagram showing what Grimaldi found in his diffraction experiment: the gap AB is the hole in the shutter, CD is the screen on which shadows are cast, and FE is the metal bar.

CATEGORY

Astronomy

Biology

Chemistry

Exploration

Mathematics

Medicine

Physics

Technology

NEWTON'S GENIUS: UNIVERSAL GRAVITATION, CALCULUS, AND THE NATURE OF COLOR

1665–69

Discoverer: Isaac Newton

Circumstances: An outbreak of plague forces a Cambridge undergraduate to spend the summer at home, where he ponders problems in the motion and attractions of bodies.

Consequences: Newton discovers law of gravitation and invents calculus

In a remarkably short period the twenty-four-year-old student created modern mathematics, mechanics, and optics. There is nothing remotely like it in the history of thought.

Derek Gjertsen,
The Newton Handbook
(1986)

© Alex Staroseltsev | Shutterstock.com

Isaac Newton's discovery of gravity (more specifically, his discovery of the inverse square law of gravity) has become perhaps the most famous of all scientific legends, in which the dreamy genius is literally struck with inspiration when an apple falls on his head. This is a popular extrapolation of an anecdote dispensed by Newton himself; it mainly serves to obscure and trivialize one of the great discoveries in history, itself just one among several landmark discoveries made by Newton in the space of a few years.

Explaining the magnitude and scope of Newton's genius is probably impossible. He had many features and qualities that have now come to be associated with genius, such as an obsessive temperament, an iron will, the ability to focus single-mindedly for long periods on a subject, and a general lack of interest in more mundane and worldly concerns. Yet his incipient genius was not clearly apparent in his early life; he struggled at school until motivated by competitiveness to outdo a bully, and at university he almost failed an early mathematics exam. But from an early age he had demonstrated an interest in building models and devices; he read widely on his own account, ranging far beyond the narrow and obsolete curriculum offered by Cambridge University when he arrived in 1661; and perhaps most importantly, he seems to have been naturally equipped with the skepticism that is the hallmark of true scientific thinking. In a notebook begun around this time he inscribed the epigraph: *"Amicus Plato amicus Aristoteles magis amica veritas."* "Plato is my friend, Aristotle is my friend, but truth is my greater friend." This was an adaptation of a phrase of Aristotle's, who had said, "Plato is my friend, but truth my greater friend." Newton was stating his commitment to the new philosophical spirit of inquiry. Above all, he was aware of the importance of experiments. "The nature of things is more securely and naturally deduced from their operations one upon another than upon the senses," he wrote in his notebook, "and when by the former experiments we have found the nature of bodies... we may more clearly find the nature of the senses."

Perhaps with this in mind, in 1664 Newton bought a small glass prism from a fair and started a series of experiments in optics. Some

TRUTH IS MY FRIEND

ISAAC NEWTON
Portait of Isaac Newton in his later years, by which time he was a distinguished scientist and civil servant, and president of the Royal Society. His great discoveries had come many years earlier, when he was little older than a teenager.

of his experiments were extremely dangerous. On one occasion he deliberately stared at the Sun to see if "my Phantasie & the Sun had the same operation upon the spirits in my optick nerve & that the same motions are caused in my braines by both." Predictably, he almost went blind and "to recover the use of my eyes shut myself up in my chamber made dark for three days together." On another occasion he wanted to test the effects on vision of compression of the eyeball, and so slid a thin knife "betwixt my eye and the bone as near to the backside of my eye as I could."

More profitable were his trials with the prism, carried out sometime between 1665 and 1669. With what he would later describe as an *Experimentum Crucis* (a "signpost" experiment), Newton discovered that white light is composed of light of different colors. It had long been known that a prism can refract white light into a rainbow, but it had been supposed that this was a property of the prism. Newton used a prism to split a ray of white light into a rainbow, and then used a pinhole in a card to isolate one of the colored rays, which he then passed through a second prism. The colored beam of light was refracted but did not change color. To complete the experiment, Newton used a lens to focus different color beams of light, recombining them to give white light.

THE PLAGUE YEARS

Far greater discoveries were to follow in an 18-month period spanning 1665–66. Newton at this time was giving much thought to some of the current problems in math and cosmology, such as the mathematics of curves and by extension the mathematics of moving bodies; the subject of gravity, the force that made things tend to Earth; and the laws governing planetary motion. In 1665 plague struck London and spread around the country, and the university was closed, with all the students sent home. Newton returned to his home at Woolsthorpe, and spent the summer doing nothing but thinking. "In those days," he would later recall, "I was in the prime of my age for invention, and minded mathematics and philosophy more than at any time since."

First, Newton invented calculus, the system of mathematical operations for working out the gradients of curved lines and the areas under curves. Since the ancient Greeks, mathematicians had tried to perform these operations using approximations based on treating a curved line as a series of straight lines (tangents). Newton knew that the French mathematician Fermat had developed methods for finding the gradients of specific curves; his achievement was to make the solution universal. "I had the hint of this method from Fermat's way of drawing tangents and by applying it to abstract equations, directly and invertedly, I made it general," he later wrote. He developed techniques for differentiation—finding the gradient of a curve—and integration—finding the area under a curve. In particular, Newton grasped the fundamental theorem of calculus, which is that differentiation is the inverse of integration. Newton used the terms fluents and fluxions; the terms used today were devised by the German polymath Gottfried von Leibniz, who discovered calculus independently in 1675 and was thus plunged into a titanic and acrimonious dispute over precedence with Newton and his supporters.

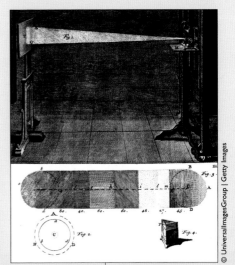

SIGNPOST EXPERIMENT

A version of Newton's *Experimentum Crucis*, showing that white light consists of different colored rays, each of which refracts to a differing degree on passing through a prism.

Newton's system of fluxions, as he described his calculus, equipped him with the mathematical tools he needed to make his most celebrated discovery. Although Newton is often said to have "discovered gravity," in practice gravity was evident to anyone who dropped an apple or fell off a wall. What Newton discovered was a quantifiable, mathematically exact law of nature describing a fundamental force that operates in both the terrestrial and celestial spheres: a law of universal gravitation.

The story of the apple came from his own account. According to the version collected by John Conduitt, husband of Newton's niece and one of his carers in his twilight years, "Whilst he was musing in a garden it came into his thought that the power of gravity (which brought an apple to the ground) was not limited to a certain distance from the earth but that this power must extend much farther than was usually thought. Why not as high as the moon said he to himself…" Today visitors to Woolsthorpe can see an apple tree said to be a descendant of the very tree observed by Newton.

Thinking about "gravity extending to the orb of the moon," as he later wrote, led Newton to wonder what must be the strength of the force of attraction that kept it in its orbit, and could this be the same as the force of gravity that caused a falling apple to accelerate to Earth? Using his newly invented calculus, Newton was able "to estimate the force with which a globe revolving within a sphere presses the surface of the sphere..."—in other words, the force with which the moon was endeavoring to fly away from the Earth, and thus the force of attraction keeping it from doing so. He found his initial comparison of the forces involved to "answer pretty well," and from consideration of Kepler's Third Law of planetary motion (see page 68) he "deduced that the forces which keep the planets in their orbs must be reciprocally as the squares of their distances from the centers about which they revolve." This is the inverse square law of gravity, which says that the gravitational attraction between two objects varies according to the square of the distance between them divided into 1 (another term used in place of "inverse" is "reciprocal"; in mathematical notation this would be written as $1/x^2$ or x^{-2}). So, a planet twice as far from the Sun as the Earth will experience ¼ of the gravitational attraction. But this law does not apply only to the planets; it holds true for the earthly sphere, with the same mathematics governing the force of gravity making an apple fall.

Among the most obvious consequences of this discovery is the technology of rockets and space travel, but more generally it ushered in a new era of science. By discovering a quantifiable, mathematically rigorous law of nature, Newton consigned huge swathes of premodern thinking to the trash can of history. His other discoveries pointed the way to a proper experimental science, laying the foundation stones for all science and technology that would follow.

A WONDERFUL SPECTACLE: VAN LEEUWENHOEK'S MICROSCOPIC WORLD

1675

CATEGORY

Astronomy

Biology

Chemistry

Exploration

Mathematics

Medicine

Physics

Technology

Discoverer: Antony van Leeuwenhoek

Circumstances: Inspired by a book on microscopy, a minor civic official from Delft constructs a simple but powerful microscope and uses it to look at a sample of lake water

Consequences: Reveals the existence of a previously invisible world of organisms

Some of these are so exceedingly small that millions of millions might be contained in a single drop of water. I was much surprised at this wonderful spectacle, having never seen any living creature comparable to those for smallness...

From a letter by Antony van Leeuwenhoek

Using a tiny lens smaller than a pea, Antony van Leeuwenhoek discovered a new world of greater magnitude and significance than any ocean-going explorer, and he did it without leaving home. His achievements were so amazing and prolific that even to this day they are met with skepticism. Though he did not invent the microscope he was, according to research scientist Brian J. Ford, one of the leading experts on Leeuwenhoek, "the father of high-power microscopy and the progenitor of microbiology."

A MOST INGENIOUS PERSON

Born humble Leeuwenhoek in 1632 (the "van" was not acquired until 1686, when his discoveries had secured him international celebrity and royal favor), Antony did not begin his remarkable researches until he was nearly 40 years old. Trained as a draper, he visited London in 1668 and came across a remarkable book that was the toast of the town: *Micrographia*, by Robert Hooke, probably the world's first professional scientist. Hooke was employed by the new Royal Society, a learned body in London devoted to the new natural philosophy, to give microscopy demonstrations, and his landmark book, with glorious plates showing familiar objects and organisms in microscopic detail, became a sensation and a bestseller when it was first published in 1665. Hooke had observed, recorded with beautiful draughtsmanship, and reproduced in the pages of his book, sights such as the hairy limbs of a flea and the rows of tiny chambers within a slice of cork, naming the latter "cells" after the bare rooms inhabited by monks. His illustrations included investigations of fabrics, and perhaps it was these that particularly caught the eye of the visiting draper, himself well used to judging the quality and value of cloth through close examination with a magnifying glass.

A LATE BLOOMER
Van Leeuwenhoek in middle age; remarkably, he was just at the start of his long and prolific career in science.

Hooke's book also included, in the preface, instructions for making a simple microscope. This was not the microscope primarily used by Hooke himself, who employed a complex and expensive instrument called a compound microscope, which used two lenses. Though easier to use, thanks to its design, the compound microscope at this time was severely limited by the poor quality of lenses available, which introduced color and focus aberrations. Since the compound microscope used two lenses, their flaws were also compounded, and the best magnification available to Hooke with his devices was around 20–30 times. He probably also used a high power single lens to help him fill in details, as evidenced by the precision and resolution of his illustrations, but apparently found it too difficult for use as his main instrument.

Leeuwenhoek had little choice in the matter — expensive and complex compound microscopes were beyond his meager resources as an amateur investigator. But through hard work he became extremely skilled at grinding his own tiny lenses, about the same size as a pin head, which he sandwiched between two metal plates with tiny apertures cut into them, and to which in turn he fixed screws that moved a pin up and down and back and forth. The specimen would be placed on the tip of the pin, and the screws could move it into focus. It was rudimentary but effective; Leeuwenhoek eventually made hundreds of them, but thanks to his skill at lens-making he could achieve astonishing magnifications of up to 300 times.

His achievement was announced to the Royal Society in April 1673 in a letter from the Dutch physician Reinier de Graaf, who reported that "a certain most ingenious person here named Leeuwenhoek has devised microscopes which far surpass those which we have hitherto seen." In August another letter from Constantijn Huygens described "our honest citizen Leeuwenhook… a diligent searcher."

Leeuwenhoek began his microscopical investigations by checking some of Hooke's observations from the *Micrographia* — of mold, a bee, and a louse, and then of cork and the shaft of a feather — some of which he found wanting. Like Hooke and others, including Italian anatomist Marcello Malpighi and Dutchman Jan Swammerdam, who had thus far looked through a microscope, Leeuwenhoek was using his microscope to look at familiar subjects. His great discovery came when he turned his gaze on something seemingly unworthy of investigation.

In 1674 he was on a boat crossing a lake called Berkelse Mere. Looking at the water he considered the way it changed in quality and opacity with the seasons. The locals told him the development of green and white growths in the summer months was related to the action of the morning dew, but Leeuwenhoek was not convinced. "I took up a little water in

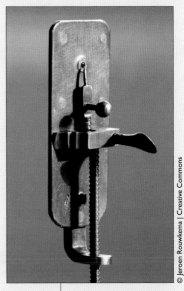

© Jeroen Rouwkema | Creative Commons

EARLY MICROSCOPE
One of van Leeuwenhoek's rudimentary but powerful microscopes. The tiny lens is the glassy dot in the circle near the top of the plate; specimens were placed on the tip of the pin in front of it.

WONDERFUL TO SEE

a glass phial," he reported to the Royal Society in a momentous letter of September 7th, 1674, "and examining it next day I found floating therein diverse earthy particles…" Looking closer, he was amazed to see "little creatures":

> … the motion of most of these was so swift, and so various—upward, downward, and round about, that 'twas wonderful to see. And I judge that some of these little creatures were more than a thousand times smaller than the smallest ones I have ever seen upon the rind of cheese.

Leeuwenhoek had discovered microbes—a catch-all term for organisms so small that they are barely visible or invisible to the naked eye. His freshwater sample contained algae, protozoans, and bacteria, kingdoms of organisms that constitute the vast majority of living things on Earth, yet which were unknown until this moment. He went on to investigate microbial populations in many other specimens, detailing them with a breathless enthusiasm that captures the excitement of discovery. For instance, describing a protozoan in 1675 he wrote: "They sometimes stuck out two little horns, which were continually moved in the manner of a horse's ears… it had a tail, nearly four times as long as the whole body, and looking as thick when viewed with my microscope, as a spider's web… I have seen several hundred little creatures, caught fast by one another in a few filaments, lying within the compass of a grain of coarse sand."

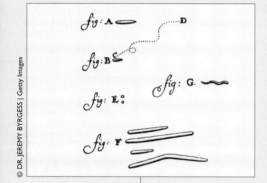

ANIMACULES
Drawings from one of van Leeuwenhoek's letters to the Royal Society, showing "animacules" observed in his own dental plaque, constituting one of the first ever observations of bacteria.

Leeuwenhoek employed a draughtsman, or limner, to make his drawings for him. On one occasion he recorded the delighted exclamation of the limner on viewing rotifers (tiny, wheel-like animals): "Suddenly there came out its roundness two little wheels, which displayed a swift rotation. The draughtsman, seeing the wheels go round and round… could not have enough of looking at them, exclaiming, 'Oh, that one could ever depict so wonderful a motion!'"

A CRAVING AFTER KNOWLEDGE

Other famous observations made by Leeuwenhoek include the first description of insect, human, and canine spermatozoa, bacteria from human dental plaque, and blood cells moving through a capillary. He became highly skilled at preparing specimens, and at very fine dissection. In 1700, for instance, he was the first to prove that what had been called "king bees" are in fact queens, when he extracted eggs from a queen honey bee.

Although he never left Delft, his letters ensured that his reputation grew and spread. In 1680 he was elected a Fellow of the Royal Society, despite never visiting, and he entertained a stream of visitors, including Queen Mary of England and, in 1698, the Russian tsar Peter the Great, who was visiting Holland to learn about ship building. Van Leeuwenhoek, as he was now known, showed him the circulation of red blood cells through the capillaries of an eel. He acquired a reputation for brusqueness, turning away even high-ranking visitors if he was not in the mood. Despite starting his research late in life he enjoyed a long and prolific career thanks to his great longevity, and even on his deathbed at the age of 90 he was still dictating the results of his final investigations, a study of gold-bearing sands for the East India Company. He even found time to describe his own respiratory condition (diaphragmatic flutter) in his final days, and did such a good job that it is still known as Leeuwenhoek's Syndrome. "My work, which I've done for a long time, was not pursued in order to gain the praise I now enjoy," he wrote in a letter of June 12th, 1716, "but chiefly from a craving after knowledge, which I notice resides in me more than in most other men. And therewithal, whenever I found out anything remarkable, I have thought it my duty to put down my discovery on paper, so that all ingenious people might be informed thereof."

CATEGORY

Astronomy

Biology

Chemistry

Exploration

Mathematics

Medicine

Physics

Technology

THE GREAT SOUTHERN LAND: JAMES COOK LANDS IN AUSTRALIA
1770

Discoverer: James Cook

Circumstances: James Cook sails the *Endeavour* to Tahiti to observe the transit of Venus, and then opens his sealed, secret orders to find the great southern land

Consequences: Charts New Zealand and eastern Australia, claiming the latter for Britain

Was it not from the pleasure which Naturly results to a man from his being the first discoverer... this kind of Service would be insupportable, especially in far distant parts like this, short of Provisions and almost every other necessary. People will hardly admit of an excuse for a Man leaving a Coast unexplored he has once discovered.

Cook's journal, Thursday, August 16th, 1770

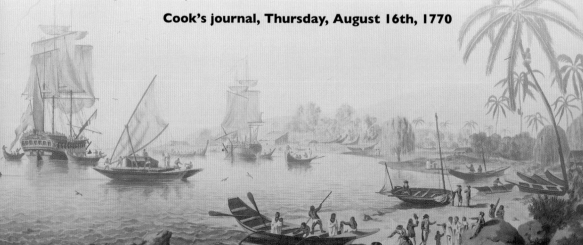

Cook's voyages of discovery constitute perhaps the greatest feats of exploration ever, combining brilliant seamanship, scientific precision, anthropological curiosity, and genuine humanity; yet he was not the first to lay eyes on any of the lands associated with him.

"What we have as yet seen of this land appears rather low, and not very hilly, the face of the Country green and Woody, but the Sea shore is all a white Sand." So wrote Lieutenant James Cook, commander of HMS *Endeavour*, on Thursday, April 19th, 1770, the day on which he first sighted Australia. Cook sailed north along the coast for nine days before attempting a landing, only to be turned away by heavy surf. The following day, April 29th, a young midshipman named Isaac Smith became the first European to set foot on eastern Australia. "Jump out, Isaac," called Cook, as the ship's boat reached the shore of what would become known as Botany Bay. But this fateful landing was attended by ominous portents of what lay in store for the true discoverers of the continent, the Aborigines, as a pair of native inhabitants "came to oppose us," hurling missiles. Cook fired two muskets at them, and, he recorded in his journal, "immediately after this we landed, which we had no sooner done than they throw'd 2 darts at us; this obliged me to fire a third shott, soon after which they both made off…"

What was Cook, a man from northwest England, doing on the other side of the world? Born in Yorkshire in 1728, Cook had worked in a grocery and haberdashery store in a fishing village, where mariners' tales filled him with a desire for a life at sea. He diligently studied mathematics and astronomy to fit himself for the tasks of navigation at which he would one day excel, and served his apprenticeship on board a Whitby collier, a stout coal-hauling ship. Transferring to the Royal Navy in 1755, Cook rose through the ranks to become a sailing master (ship's navigator), making his name by preparing charts of the St. Lawrence waterway in Quebec, which would directly contribute to British conquest of the region. Having further honed his navigation and cartography skills in a survey of the Newfoundland coast, in 1768 he was selected by the Royal Navy to lead an exciting new mission jointly sponsored by the Royal Society.

THE FIRST VOYAGE

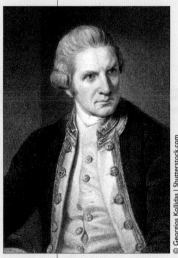

© Georgios Kollidas | Shutterstock.com

RISEN FROM THE RANKS
Cook had risen from the ranks to become an officer, and he took the care of the men under his command seriously; he was particularly proud of his success in keeping his crew free from scurvy.

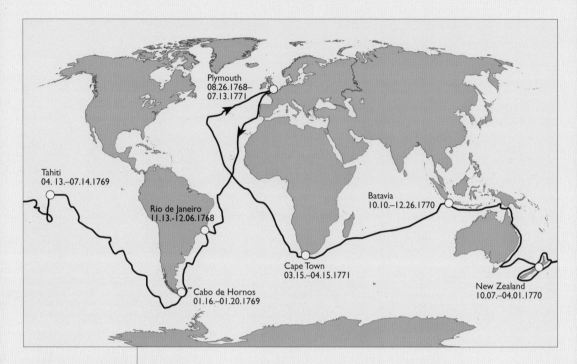

Plymouth
08.26.1768–
07.13.1771

Tahiti
04. 13.–07.14.1769

Rio de Janeiro
11.13.-12.06.1768

Batavia
10.10.–12.26.1770

Cape Town
03.15.–04.15.1771

Cabo de Hornos
01.16.–01.20.1769

New Zealand
10.07.–04.01.1770

**THE FIRST VOYAGE
OF CAPTAIN COOK**
Cook's epic voyage was
just the first of three,
in the course of which
Cook would sail well over
100,000 miles and lay the
basis for British colonization
of Oceania.

Commissioned as a lieutenant and given command of an adapted Whitby collier, HMS *Endeavour* — a sturdy, shallow-drafted vessel that could cope with tough conditions, approach close to coastlines, and carry extensive stores for long voyages, yet which could be managed with a reduced crew if necessary — Cook was tasked with sailing to the other side of the world in the name of science. Astronomers wanted to measure interplanetary distances by observing the transit of Venus across the Sun, scheduled to take place on June 3rd, 1769, but to do so they would need observations taken from locations as far apart as possible. Hence, the *Endeavour* was sent to Tahiti, in the South Pacific, with members of the Royal Society on board, notably an astronomer, Dr. Green, and a wealthy young botanist, Joseph Banks, who was accompanied by artists and a Swedish naturalist, Dr. Daniel Solander.

The expedition set off from Plymouth on August 25th, 1768, reaching Tahiti by way of Cape Horn on April 13th, 1769. They successfully tracked the transit of Venus: "we had every advantage we could desire in observing the whole of the planet Venus over the sun's disk. We very distinctly saw an atmosphere or dusky shade around the body of the planet." Then Cook opened sealed, secret Admiralty orders commanding him to proceed south and west to look for the *Terra Australis Nondum Plene Cognita* — "the southern land not yet known" — aka the Great Southern Land. Geographers had long supposed that there must be a vast continent at the Antipodes to balance the great land masses of the northern hemisphere, and hints of its existence had been picked up by the Dutch explorer Abel Tasman over a century before Cook, when he sighted what is now Tasmania and approached the coast of New Zealand before being chased away by native war canoes.

In October the *Endeavour* reached New Zealand, encountered terrifying cannibals and charted the coastline, establishing that it was two islands and not part of any putative southern continent. Striking westward in March 1770, he reached the coast of what he came to call New South Wales, and finally made landfall on April 29th. "The great quantity of Plants Mr. Banks & Dr. Solander found in this place," he recorded in his journal, "occasioned my giving it the Name of Botany Bay; it is situated in the Lat.de of 34.'0 So. Long.de 208:37 W.t; it is Capaceous safe & Commodious..."

The most dangerous passage of the journey was to follow. In June the *Endeavour* was nearly wrecked on the Great Barrier Reef, taking seven weeks to repair, after which they carefully threaded their way north. Cook was determined to see if the northern coast of Australia was connected to New Guinea. Having proved that it was not, he sailed into Batavia in the Dutch East Indies, having lost remarkably few of his crew to illness thanks to his diligent precautions against scurvy. Alas, many now took ill and perished with dysentery, before the *Endeavour* returned home via South Africa, reaching England in July 1771.

On returning to England, Joseph Banks claimed to have been the star of the show, and indeed he had made many extraordinary discoveries in botany and zoology. But Cook's reputation grew on the publication of his journals, with widespread appreciation for his feats of navigation and cartography. In 1772, now a captain, he led a second expedition to the South Pacific equipped with two ships, the *Resolution* and the *Adventure*, to settle once and for all the question of the southern continent. Crossing the Antarctic Circle, Cook and his crew ventured farther south than anyone before, but never got within sight of Antarctica itself:

THE SECOND AND THIRD VOYAGES

The outer or northern edge of [an] immense ice field was composed of loose or broken ice so close packed together that nothing could enter it. About a mile in began the firm ice, in one compact solid body and [it] seemed to increase in height as you travel it to the south. In this field we counted ninety-seven ice hills or mountains, many of them vastly large.

GRISLY BUSINESS
An illustration from an 1815 edition of Cook's journal, showing him witnessing a human sacrifice in Otaheite (Tahiti).

Covering approximately 70,000 miles (110,000 kilometers) — the equivalent of circumnavigating the world three times — this second voyage finally proved that there was no Great Southern Land in the classical sense, as well as discovering several Pacific islands and mapping Easter Island.

In 1776 Cook, now a fellow of the Royal Society, led a third expedition to search for the Pacific gateway to the Northwest Passage — a maritime route across the top of North America. Charting the northwest coast of North America he was eventually driven back by sea ice, returning to Hawaii, where he met with tragedy. Tensions between the natives and Europeans over bad behavior on both sides led to a skirmish on February 14th, 1779, at Kealakekua Bay. Cook blasted a Hawaiian with his shotgun but was fatally stabbed, and his body was roasted in a pit for six hours. Some of his remains were recovered and buried at sea in a tiny coffin, weighted with cannonballs.

His legacy included invaluable and highly accurate charts, a new, more scientific approach to exploration and discovery, and the opening of new lands to European settlement. In 1788, less than a decade after Cook's death, the First Fleet sailed into Botany Bay to establish the first British colony in Australia, with devastating consequences for the Aborigines, who had arrived at least 50,000 years earlier. New Zealand too saw the majority of the indigenous population killed off by disease and war, with European colonization in their place; the entire country was annexed to Great Britain in 1840. Cook himself is remembered and celebrated today as one of the greatest explorers of all time — a man who went to the ends of the earth, very much as he saw himself: "My ambition leads me not only farther than any man has gone before me, but as far as I think it possible for man to go."

OXYGEN: JOSEPH PRIESTLEY AND THE BREATH OF LIFE

1774

CATEGORY

Astronomy

Biology

Chemistry

Exploration

Mathematics

Medicine

Physics

Technology

Discoverer: Joseph Priestley

Circumstances: A self-taught radical and scientist in Birmingham aims a burning glass at a heap of red calx of mercury

Consequences: Discovers oxygen and the principle behind respiration and combustion

Had it been common air, a full-grown mouse [...] would have lived in it about a quarter of an hour. In this air, however, my mouse lived a full hour...

Joseph Priestley, *Experiments and Observations on Different Kinds of Air*, Vol II (1776)

Joseph Priestley and his mice opened the door to a new chapter in science that would become known as the Chemical Revolution, revealing the secret substance that constituted the breath of life and the spirit of fire. Yet Priestley and his rival Antoine Lavoisier, the great French chemist, would fight bitterly over who should take precedence as the discoverer of oxygen, before both men were swept up in a revolution of a different kind.

A SERVICE TO THIRSTY SOULS

If the 17th century had been the century of physics, with incredible discoveries in astronomy and optics, the 18th century was surely the century of chemistry. New compounds and elements were identified at an accelerating rate, and in the years 1750–80 the primary focus was on gases, although the concept of a "gas" was still forming, and researchers referred to "airs" of different types. In this period, observed the historian of science J. L. Heilbron, "Novel airs then began to rise promiscuously."

In 1754–56 the Scottish scientist Joseph Black had identified a gas that he called "fixed air," because until released by heating it seemed to be fixed as part of a solid, which is now known as carbon dioxide. Black showed his fixed air to be the cause of effervescence when limestone is added to an acid. Black's discoveries inspired the eccentric English scientist Henry Cavendish—a man so pathologically shy that he would only communicate with his housekeeper by leaving notes, and who had a separate staircase built for his servants so that he wouldn't bump into one—to pneumatic investigations of his own. Cavendish discovered what he called "inflammable air," now known as hydrogen, and in 1781 disproved the ancient notion that water is an element by combining inflammable and common air and lighting them with an electric spark to produce water. Lavoisier would later prove that the hydrogen gas combines with oxygen in the "common air" to produce water (aka dihydrogen oxide, or H_2O).

DISSENTING VOICE
Joseph Priestley, Unitarian preacher and pneumatic scientist, whose support for the revolutions abroad aroused the anger of the mob at home.

Also following up on Black's fixed air was Joseph Priestley, who would discover no less than eight new airs of his own. Priestley came from a humble background and held strong but radical religious and political views. Largely self-taught, in 1766 Priestley met and befriended the American scientist Benjamin Franklin, who encouraged him to pursue natural philosophy. Shortly thereafter, Priestley got a job as a minister in Leeds and moved into a house next door to a brewery, where the fermentation vats provided a rich source of Black's fixed air. He began to experiment with the gas, and managed to recreate the natural effervescence of some spring waters by forcing fixed air into water under pressure, thus creating soda water. Priestley refused to patent his invention, making it available to all as a potential health benefit, "a service to naturally, and still more to artificially, thirsty souls," wrote the scientist and science popularizer Thomas Huxley in an essay of 1874, "which those whose parched throats and hot heads are cooled by morning draughts of that beverage, cannot too gratefully acknowledge."

In August 1771 Priestley made the first steps in the discovery of photosynthesis. It was known that placing a burning candle or an animal in an enclosed space such as a glass dome exhausted some vital principle essential for both combustion—so that the candle flame would gutter and die—and life—so that a mouse would perish. Priestley placed a burning candle alongside a sprig of mint into a sealed glass-walled space. The candle soon went out, but after 27 days Priestley used his burning glass—a magnifying lens or parabolic mirror that could focus the rays of the sun to produce intense localized heating—to relight it, observing that the candle now burned well in air previously unable to support combustion. He concluded that the green plant had somehow restored the vital air necessary for combustion and/or respiration; perhaps, he speculated, "the injury which is continually done [to the ability of the air to support respiration] by such a large number of animals is, in part at least, repaired by the vegetable creation." Priestley had grasped at least one aspect of the process of photosynthesis, in which green plants create oxygen and replenish atmospheric stocks used up by other processes, such as animal respiration.

Exactly three years later he performed his most celebrated experiment, using a large burning lens to heat red calx of mercury (now known to be mercury (II) oxide) inside a sealed glass container, and obtaining large amounts of an "air." In his most important book, *Experiments and Observations on Different Kinds of Air*, he relates: "I presently found that, by means of this lens, air was expelled from [the calx] very readily… what surprised me more than I can well express, was that a candle

SUPERIOR AIR

burned in this air with a remarkably vigorous flame… a piece of red-hot wood sparkled in it, exactly like paper dipped in a solution of niter, and it consumed very fast." This air was, he wrote, "five or six times as good as common air." He found that the air could keep a mouse alive for four times longer than a similar quantity of "common air" and that when he breathed it himself, "I fancied that my breast felt peculiarly light and easy for some time afterwards. Who can tell but that in time, this pure air may become a fashionable article in luxury. Hitherto only two mice and myself have had the privilege of breathing it."

THE PHLOGISTON DEBATE

Priestley had discovered oxygen, which he called "dephlogisticated air," interpreting his finding according to the dominant theory of the day. The phlogiston theory held that burning a material caused it to give off phlogiston, the principle of combustibility. Priestley theorized that since his "superior air" supported burning so well, it must be particularly good at absorbing phlogiston, suggesting in turn that it was initially devoid of the stuff, hence "dephlogisticated." Visiting Paris in 1774, Priestley related his findings to Lavoisier, who repeated his experiments but drew quite different conclusions. The phlogiston theory was already struggling to cope with mounting evidence against it, specifically the finding that burning a solid in air causes the solid to gain mass while the air loses mass. Lavoisier realized that the mass in question was Priestley's new air, and that the phlogiston doctrine was back to front. Testing a variety of acids, Lavoisier found Priestley's dephlogisticated air to be present in all of them, and so he renamed it oxygen, from the Greek for "acid forming." He was able to present a fully formed "general theory of combustion" far superior to the now destroyed phlogiston doctrine: combustion (burning) and respiration (breathing) involve not giving *off* phlogiston but combining with oxygen. Burning charcoal with oxygen produces Black's "fixed air" (carbon dioxide). Combining oxygen with Cavendish's "inflammable air" produces water, leading Lavoisier to rename "inflammable air" as hydrogen (Greek for "water maker").

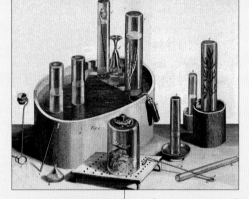

PNEUMATIC EQUIPMENT
The pneumatic trough that Priestley used to isolate and collect gases, together with other equipment, including a bell jar occupied by a mouse.

Lavoisier now claimed to be the discoverer of oxygen, prompting a precedence dispute with Priestley. Both men overlooked the claims of the Swedish chemist Karl Scheele, who had actually discovered what he called "empyreal air" ("fire air") prior to 1771. Unfortunately for Scheele, he only got round to writing up his discovery in 1773 and then had to wait a further four years for his mentor to finish writing the preface, so that his *Chemical Treatise on Air and Fire* was not published until 1777, three years after Priestley.

In 1789 the Revolution convulsed France and appalled Britain. Priestley, associated with Revolutionary sentiment thanks to his radical politics and religion, became the target of an anti-Revolutionary mob and was driven out of Birmingham, as Huxley dryly related:

> In 1791, the celebration of the second anniversary of the taking of the Bastille by a public dinner, with which Priestley had nothing whatever to do, gave the signal to the loyal and pious mob, who… had the town at their mercy for three days… Priestley and his family had to fly for their lives, leaving library, apparatus, papers, and all their possessions, a prey to the flames… Priestley never returned to Birmingham.

Priestley fled to London but found the scientific and academic establishment hostile, and was forced to move to America, where he lived out his days in obscurity, still clinging to the phlogiston doctrine. Lavoisier fared even worse; having been involved with tax collectors and an unpopular anti-smuggling campaign, he was sent to the guillotine in 1794. The eminent mathematician Joseph Lagrange famously lamented, "It took only a moment to cut off that head, yet a hundred years may not give us another like it."

CATEGORY

~~Astronomy~~ **Astronomy**

Biology

Chemistry

Exploration

Mathematics

Medicine

Physics

Technology

THE GEORGIAN PLANET: WILLIAM HERSCHEL DISCOVERS URANUS

1781

Discoverer: William Herschel

Circumstances: Amateur astronomer spies something never before seen in the heavens

Consequences: Doubles the known size of the solar system; fires popular and Romantic conceptions of astronomy and science

I had gradually perused the great Volume of the Author of Nature and was now come to the seventh Planet.

William Herschel, in a letter to Dr. Hutton (1809)

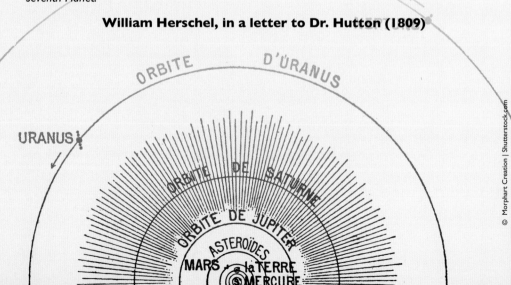

"Of all the discoveries in this science," wrote John Bonnycastle in his bestselling 1786 popular guide *An Introduction to Astronomy: In a Series of Letters from a Preceptor to his Pupil*, "none will be thought more singular than that which has lately been made by Dr. Herschell... a Primary Planet belonging to the solar system, which till 13th of March 1781... had escaped the observation of every other astronomer, both ancient and modern." Bonnycastle was referring to Uranus, the first new planet to be discovered for over one and half a thousand years, since the time of Ptolemy.

Almost 5 billion years ago a vast cloud of dust and gas in interstellar space was disturbed, possibly by shockwaves from a nearby supernova, causing one region to become denser than the others. This quickly sucked in most of the cloud to become a protostar, and the gravitational attraction of this star compressed the rest of the matter into an "accretion disc." This was how our solar system began. Because the proto-sun at the center of the system was not too big or too hot, it did not sweep up or vaporize the accretion disc that formed around it, and the disc grew cold enough for some of the gas to freeze into ice particles, which began to accrete into asteroid-like "planetismals" as they collided with and attracted one another. Around 4.6 billion years ago the system settled down into a relatively stable equilibrium, by which time there were eight planets and a large number of "left-over" asteroids and planetismals (including the object known as Pluto). The third planet from the Sun is Earth, and five of the other planets are close enough to the Sun for an observer on this planet to see them with the naked eye. Thus it has been known since prehistoric times that there are five celestial bodies that do not behave like the other stars; they are not fixed in the sky, as the other stars seem to be, but change position in a way that clearly shows they are orbiting some center of attraction: they are the planets Mercury, Venus, Mars, Jupiter, and Saturn. But in addition to these there are two more that orbit the Sun at such colossal distances that they are invisible without powerful telescopes of great magnifying power, which did not exist until a musician and amateur astronomer named William Herschel built them in the late 18th century.

THE FORMATION OF THE SOLAR SYSTEM

© Georgios Kollidas | Shutterstock.com

FROM HANOVER TO WINDSOR
William Herschel, whose career traced a curious trajectory from teenage regimental oboist in Hanover, to professional musician in Bath, to Court Astronomer in Windsor.

Born Friedrich Wilhelm Herschel in 1738, William Herschel, as he was to become known in Britain, started life in the German state of Hanover, which had recently supplied Britain with a new royal family. As a young teenager he followed his father into musical military service as an oboe player in a regimental band, but did not fare well in warfare and moved to Bath in the west of England to pursue a career in music, as well as his private passion, stargazing. Teaching himself to make reflecting telescopes, where a mirror is used to gather light and focus it through eyepieces, and, in 1772, recruiting his younger sister Caroline to the cause, Herschel underwent extraordinary labors of casting and polishing metal mirrors to produce the most powerful telescopes then available. By 1781 he and Caroline were engaged in an exhaustive project to catalog every double star in the sky, and in March of that year he happened to be gazing at a particular patch of sky with his high magnification instrument.

THE MOST LUCKY OF ASTRONOMERS

On Tuesday, March 13th, Herschel recorded in his "Observation Book" the following entry: "In the quartile near Zeta Turi… is a curious either nebulous star or perhaps a Comet." In fact, he suspected quite early on that this unexpected object was neither a star nor a comet. If it were a star it would appear "fixt"—in other words, it would not change position night by night; Herschel soon established that it did, and he even thought that he saw it growing larger. If it were a comet it should display the characteristic hazy "beard" or "tail," caused, it is now known, by the solar wind blasting tiny particles off the lump of dirty ice that comprises a comet. On April 6th he noted: "I viewed the Comet with 460 [magnifications] pretty well defined, no appearance of any beard or tail."

By this time Herschel had already spread news of his discovery and clearly implied his inference that it was a new planet. Apprised of Herschel's discovery, the Astronomer Royal Nevil Maskelyne cautiously confirmed it, writing to the amateur astronomer on April 23rd:

> Sir, I am to acknowledge my obligation to you for the communication of your discovery of the present Comet, or planet. I don't know which to call it. It is as likely to be a regular planet moving in an orbit nearly circular round the sun, as a Comet moving in a very eccentric ellipsis… On the 6th April I viewed the Comet [with the greatest magnification] and saw it a very sensible size… This […] showed it to be a planet and not a fixt star.

Maskelyne henceforth supported Herschel in his claims to the Royal Society, although he also told the society's president, Joseph Banks, that "Mr. Herschel is undoubtedly the most lucky of Astronomers in looking accidentally at the fixt stars with a 7-foot reflecting telescope magnifying 227 times to discover a comet of [tiny] diameter, which if he had magnified only 100 times he could not have known from a fixt star… Perhaps accident may do more for us than design could…" The implication that Herschel was merely lucky would dog him.

The discovery of the new planet was confirmed by a stream of astronomers from across Europe. In October the Russian mathematician Anders Lexell wrote from St. Petersburg, revealing that he had calculated the orbit of the new planet—it was huge and very, very far away, around twice as far from the Sun as the next nearest planet, Saturn. In fact, Uranus travels up to 20 times farther away from the Sun than the Earth. Herschel's discovery had doubled the size of the Solar System at a stroke.

OF HIS OWN DESIGN
A copy of the 7-foot (2-meter) reflecting telescope designed and built by Herschel himself, with which he discovered Uranus.

SOME WATCHER OF THE SKIES

For Herschel, the discovery was transformative. Brought to London to dine with the Astronomer Royal and the President of the Royal Society, he was feted and assured that he would soon be elected a Fellow and awarded the Copley Gold Medal. The British scientific establishment was delighted to have stolen a march on the French astronomers who had hitherto dominated the field, and Banks humorously encouraged Herschel to provide the new planet with a name "forthwith… or our nimble neighbors, the French, will certainly save us the trouble of Baptizing it." Taking advice from his friend Dr. William Watson, Jr, Herschel diplomatically suggested it be named after his Hanoverian compatriot, the King: Georgium Sidus—the Georgian Star. Other European astronomers favored a name from mythology without political or nationalist overtones, and it was suggested that since, in Roman mythology, Saturn is the father of Jupiter, this new planet beyond Saturn should be named after the father of Saturn, Uranus, a name that chimed with the goddess of astronomy, Urania.

Herschel had to rebut sniping to the effect that his discovery was accidental. Writing to Banks in November he insisted: "The discovery cannot be said to be owing to chance only it being almost impossible that such a star should escape my notice… The first moment I directed my telescope to the new star, I saw… that it differed sufficiently from other celestial bodies, and… was quite convinced that it was not a fixt star." But also he was showered with praise. "Mein Gott! If I had only known, when I was for a few days in Bath in October 1775, that such a man lived there!" wrote the eminent German astronomer Georg Christoph Lichtenberg.

A REVOLUTION IN THE POPULAR CONCEPTION OF COSMOLOGY

According to Richard Holmes, in his book *Age of Wonder: How the Romantic Generation Discovered the Beauty and Terror of Science*, the discovery of Uranus "reignited the general fascination with astronomy [and] began a revolution in the popular conception of cosmology." It also did much to fire the Romantic imagination, enshrining the concept of the solitary genius toiling alone at his instruments to probe the mysteries of the cosmos and unlock the secrets of nature. Herschel's personification as this Romantic ideal is perhaps most clearly articulated in the famous lines from John Keats' 1816 poem "On First Looking into Chapman's Homer": "Then felt I like some watcher of the skies; When a new planet swims into his ken…" Keats had been awarded Bonnycastle's *Introduction to Astronomy* as a schoolboy, and evidently Herschel's discovery made an impression.

ANIMAL ELECTRICITY: GALVANI DETECTS THE PRESENCE OF BIOELECTRICITY

1781

CATEGORY

Astronomy

Biology

Chemistry

Exploration

Mathematics

Medicine

Physics

Technology

Discoverer: Luigi Galvani

Circumstances: Galvani experiments on the link between electricity and muscular activity in frogs' legs

Consequences: Triggers dispute over source and nature of animal electricity

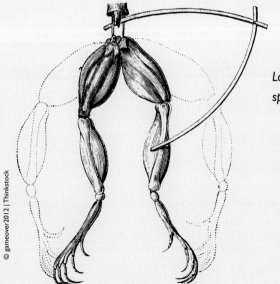

Lo and behold, the frogs began to display spontaneous, irregular, and frequent movements.

Luigi Galvani, *De Viribus Electricitatis in Motu Musculari Commentarius* (Commentary on the Effect of Electricity on Muscular Motion), (1791)

Surrounded by dissected corpses and flayed muscles, electrostatic generators and metal wires and hooks, the anatomist applies an electric shock to his dead specimen and beholds with wonder as it jerks into reanimated motion. This is not a scene from *Frankenstein*, but from real life, although the similarity is not accidental: the anatomist is Luigi Galvani, and he has just discovered the secret of life.

ANIMAL SPIRITS

An ancient Egyptian hieroglyph provides the first written record of bioelectricity—the phenomenon of electricity in living organisms—describing how the sheatfish "releases the troops." This is believed to be a reference to an electric catfish capable of generating shocks of more than 450 volts, thus forcing ancient fishermen to release their catch or risk electrocution from a single sheatfish caught in the net. Electricity remained a poorly understood and elusive phenomenon, and models of physiology had no inkling of its role in muscle action. Instead, the influential ancient physician Galen proposed that the neuromuscular mechanism involved fluid "animal spirits," a theory brought up to date in the 17th century by Descartes, who believed that the role of the nerves is to conduct fluids between the brain and the muscles, writing in his 1664 *Traite de L'Homme* (Treatise on Man):

> … one can well compare the nerves of the machine that I am describing to the tubes of the mechanisms of these fountains, its muscles and tendons to diverse other engines and springs which serve to move these mechanisms, its animal spirits to the water which drives them, of which the heart is the source and the brain's cavities the water main.

Descartes's "hydraulic theory" of neuromuscular mechanism was even backed up by anatomists, who could conceive of no other role for the brain than the "conversion of vital spirits… into essential animal spirits," in the words of 17th-century English anatomist Thomas Willis.

THE ANATOMIST
Luigi Galvani, who studied medicine at Bologna before going on to lecture on the anatomy of the frog, while simultaneously pursuing investigations into the new science of electricity.

In 1664, however, the Dutch scientist Jan Swammerdam's experiment proved Descartes to be in error. Sealing a frog muscle in a glass vessel from which projected a narrow tube holding a water droplet, Swammerdam made the muscle contract by stimulating it with a silver wire attached to a copper loop. The contraction failed to make the water droplet move, showing that the muscle had not changed its overall volume during the contraction, contradicting the hydraulic theory.

It is possible that, with this experiment, Swammerdam achieved the first neuromuscular electric stimulation, although it may also be that he mechanically stimulated the muscle. Accordingly, it is Galvani who is traditionally credited with the first demonstration of bioelectricity.

ELECTRICAL MATTERS

The invention of static electricity generators by Otto von Guericke and Francis Hauksbee, the Elder in 1672 and 1704 respectively sparked a wave of interest in electricity, leading to the invention of the Leyden jar. This device, now known as a capacitor, can store electrical charge to be dispensed at will, though only in the form of an uncontrolled, one-off shock. A Leyden jar is a glass bottle coated inside and out with metal foil, and charged by means of an electrically conducting metal pin that projects through the neck. If a metal arch is held so that one end touches the outer foil surface and the other the top of the pin, an electric shock is released, or if the external rod is held very close to the pin a spark may leap across the gap, carrying the shock. The effects of the jar were widely compared to the shock received from electrical animals such as the electric eel or the torpedo (electric ray), leading in turn to speculation that, as suggested by Albrecht von Haller, professor of anatomy and medicine at the University of Göttingen in Germany, the nature of the "nervous fluid" might be due to "electrical matter" present in the "animal spirits."

Such speculations prompted the investigations of Luigi Galvani, professor of anatomy at the world's oldest university in Bologna, Italy. He dissected frogs to leave part of the spinal cord projecting above the legs, and hooked them up to various electrostatic generators and Leyden jars by means of wires attached to the nerves and muscles, often holding the specimen in place by means of an iron hook driven through the spinal cord. On January 26th, 1781, his assistant— probably his wife Lucia—touched a nerve with a scalpel just as a nearby "electrical machine" discharged a spark. At this moment, "all the muscles of the legs seemed to contract… as if they were affected by powerful cramps." Could it be, Galvani wondered, that the "electrical atmosphere" (electrical field) caused by the spark had stimulated movement of the electrical fluid in the frog's nerves?

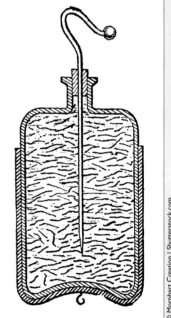

© Morphart Creation | Shutterstock.com

LEYDEN JAR
Cross-section of a Leyden jar, showing the foil-covered glass and the metal pin isolated in the neck of the jar.

© Photos.com | Thinkstock

ELECTRICAL LABORATORY
Scene similar to that presented by Galvani's laboratory, with devices for generating electrical charge and frogs' legs prepared for experimentation.

This observation prompted Galvani to undertake a systematic inquiry. Following Franklin, he wanted to see if lightning really is a form of electrical spark, and if so, whether it could stimulate muscle action. Setting brass hooks in the frogs' spinal cord to an iron railing in his garden, Galvani noticed that the frogs' legs went into contractions "not only when the lightning flashed but even at times when the sky was quiet and serene." It was this last finding that was most exciting for Galvani.

> In early September [1786], at twilight, we placed… the frogs prepared in the usual manner horizontally over the [iron] railing. Their spinal cords were pierced by iron hooks, from which they were suspended. The hooks touched the iron bar [of the railing]. And, lo and behold, the frogs began to display spontaneous, irregular, and frequent movements.

No electrical field or current was present, yet still the frog legs kicked; to Galvani it indicated that the source of the animating electrical fluid *must* be the muscles themselves. In other words, here was proof that living matter was the source of bioelectricity. The final piece of the puzzle fell into place for Galvani when he showed that he could elicit the contractions by simultaneously touching the nerve and muscle of the frog with a bimetallic arch of copper and zinc, just as one elicited a discharge from a Leyden jar.

AN INEPT HYPOTHESIS?

Aware that his conclusions would be contentious, Galvani waited until 1791 to publish his essay *"De Viribus Electricitatis in Motu Musculari Commentarius"* ("Commentary on the Effect of Electricity on Muscular Motion"). In it, Galvani explained his belief that the muscle acts as a sort of biological Leyden jar:

> It would perhaps not be an inept hypothesis and conjecture, nor altogether deviating from the truth, which should compare a muscle fiber to a small Leyden jar, or other similar electric body, charged with two opposite kinds of electricity; but should liken the nerve to the conductor, and therefore compare the whole muscle with the assemblage of Leyden jars.

He thought that the nerves conducted electrical fluid to the outer surface of the muscle, equivalent to a conductor carrying charge to the outer surface of a Leyden jar, causing the inner surface to become oppositely charged, with the opposition leading to muscular contraction.

Galvani was particularly anxious to hear the response of one of the leading experts on electricity, Alessandro Volta of the nearby University of Pavia. Though impressed by Galvani's discovery, Volta insisted that the explanation was all wrong. The muscle acted not like a Leyden jar, but like an electroscope, a primitive detector of electrical current. The electricity is not resident in the muscle and nerves, but is being generated by the contact between two metals in the bimetallic arch; the contraction of the muscle is merely an indication of the potential created by the external device. Volta's critique divided the scientific world, resulting in an intense controversy between "animalists" and "metalists." Galvani and his supporters carried out further experiments showing that biological tissue could be the source of the electrical potential; attempting to disprove them, Volta was led toward a historic discovery of his own (see page 117).

The irony is that both men were correct. Volta was correct in that the source of electricity in Galvani's experiment was indeed the bimetallic arch, and that the frog muscle was analogous not to a Leyden jar but to an electroscope. But Galvani was correct in surmising that muscular contractions are triggered by electrical impulses passing via the nerves, and that living tissue can indeed generate its own electricity, by means of building up opposing charges on the inner and outer surfaces.

Whatever the true explanation, Galvani's discovery had historic consequences for the study of electricity, chemistry, biology, and the intersections between them, making clear the connection between chemistry and electricity. Perhaps Galvani's most famous legacy, however, is fictional, not scientific. His famous experiment, with its reanimation of dead tissue achieved by the application of mysterious electricity, became one of the main inspirations for Mary Shelley's *Frankenstein*, in which the monster is brought to life by means of bioelectricity. Discussing research on electricity with Lord Byron and Percy Shelley, Mary recalled musing that "Perhaps a corpse would be reanimated; galvanism had given token of such things." These discussions would lead directly to the famous passage:

> ... I collected the instruments of life around me, that I might infuse a spark of being into the lifeless thing that lay at my feet... I saw the dull yellow eye of the creature open. It breathed hard, and a convulsive motion agitated its limbs.

CATEGORY

Astronomy

Biology

Chemistry

Exploration

Mathematics

Medicine

Physics

Technology

VACCINATION: JENNER'S MEDICAL BREAKTHROUGH

1796

Discoverer: Edward Jenner

Circumstances: When a local milkmaid presents with a case of cowpox, country doctor Edward Jenner takes the opportunity to test an old wives' tale

Consequences: Vaccination becomes a global practice; smallpox is eradicated

It is owing to your discovery that in the future the peoples of the World will learn about this disgusting smallpox disease only from ancient traditions.

Thomas Jefferson, president of the United States, in a letter to Edward Jenner, 1806

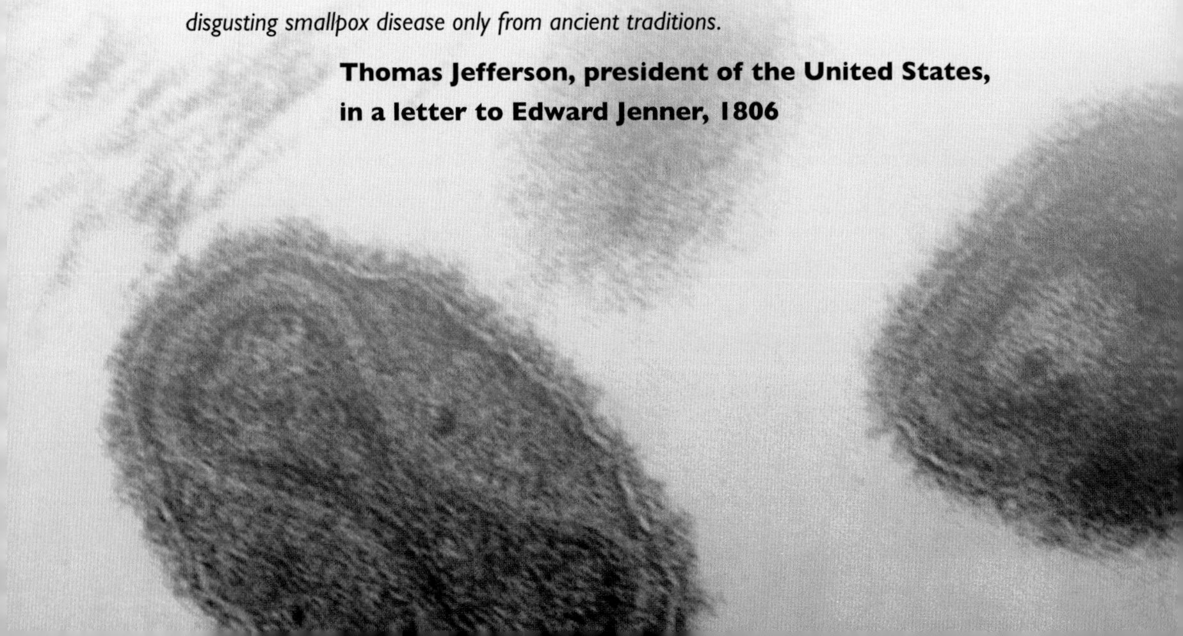

One of the greatest discoveries in the history of medicine, vaccination is the artificial stimulation of immunity to a disease. Controversial then as now, the story of Edward Jenner and vaccination sheds light on the meaning of discovery and the power of science to benefit humanity.

Smallpox was a virulent and appalling disease, present in the Old World since prehistoric times. Ancient Egyptian mummies from the 16th century BCE show evidence of the facial scarring characteristic of the disease, and it spread to India and China by the first millennium BCE, reaching Europe by at least the 5th century CE if not earlier. European explorers and colonizers introduced smallpox to the New World in the 15th century, with devastating effect. Edward Jenner called it "the speckled monster," observing that "there is no disease which presents a more melancholy scene than the natural smallpox as it frequently occurs." Caused by the variola virus, smallpox was highly contagious. The afflicted suffered from fever, nausea, and a spreading rash that resolved into hard pustules or "pox," which could join up or scab over and fall off, leaving deep, pitted scars, including of the eyes. In 18th-century Europe smallpox killed 400,000 people annually. Fatality rates varied from 20 to 60 percent, and for infants it was much higher, especially in the cities: 80 percent in London and a horrific 98 percent in Berlin during the late 1800s. Survivors were often disfigured and a third of them were left blind.

THE SPECKLED MONSTER

Once infection had taken hold there was no treatment for this horror, but by the late 1700s there was a widely practiced form of prevention, known as inoculation, from the Latin for "grafting," or variolation, from the name of the disease. Inoculation involved taking scab material from someone infected with a mild form of the disease and scratching it into the skin of a healthy person, in the hope that a mild version of the disease would be triggered that would leave the variolated subject with immunity to severe smallpox. Variolation had its own dangers: about 2–3 percent of variolated people would contract fatal smallpox, trigger fresh epidemics, or pick up other blood-borne diseases such as tuberculosis and syphilis, but this was ten times less fatal than naturally occurring smallpox.

Among those routinely variolated in Gloucestershire in 1757 was an eight-year-old boy named Edward Jenner. Jenner vividly recalled being starved, purged, and bled prior to the inoculation, before being locked in a stable with other variolated boys to let the infection take its course. Although the infection he suffered was relatively mild, the experience profoundly marked him.

A DISEASE OF SO PECULIAR NATURE

Of minor gentry, Jenner went into medicine, beginning with an apprenticeship to a country surgeon in Chipping Sodbury at age 13. It was here that he first encountered the disease cowpox, little known outside of rural cattle farming communities, and heard the common lore that dairymaids would keep their pretty faces since they were protected against smallpox, and specifically that it was cowpox infection that conferred this advantage. Cowpox is a disease that causes pustules on the udders of infected cows, and it can spread to humans by close contact; human sufferers experience mild fever and blisters on the hands and arms.

VACCINATION CHAMPION
Edward Jenner, a country physician with contacts in the intellectual milieu of the metropolis, and the scientific training to make a decisive case for the virtue of vaccination.

Jenner went on to study in London, secure a country practice at Berkeley in Gloucestershire, and pursue a range of scientific interests. He was friends with Joseph Banks, later President of the Royal Society, and even helped with the cataloging of Banks's specimens from the *Endeavour* expedition (see page 92). Jenner experimented with hydrogen balloons and studied geology. In 1789 he was made a Fellow of the Royal Society for his research on the cuckoo; he was the first to discover that the baby cuckoo pushes the chicks and eggs of its foster parents out of the nest. But he never forgot the country lore about cowpox, later writing:

In the present age of scientific investigation it is remarkable that a disease of so peculiar nature as the cow-pox, which has appeared in this and some of the neighboring counties for such a series of years, should so long have escaped particular attention.

Jenner started to collect case studies of people who had been infected with cowpox, checking to see whether they had subsequently suffered smallpox. Like several other people around this time, he was excited at the prospect that here might be a weapon against the scourge of the era. In 1765, for instance, Jenner's friend and fellow country physician John Fewster had presented to a London society a paper called "Cow pox and its ability to prevent smallpox." Jenner's great contribution was to approach the issue systematically and scientifically, having learned the great power and value of the scientific method from his mentor, the famous surgeon John Hunter.

Jenner knew that a proper scientific evaluation of the claims relating to smallpox required several steps: observations, leading to the formulation of a hypothesis, which in turn would make predictions that could be tested by experiment. He had the first two stages, and in 1796 he was presented with the opportunity to complete the process. In May of that year he was consulted by Sarah Nelmes, "a dairymaid at a farmer's near this place, [who] was infected with the cow-pox from her master's cows." Specifically, Nelmes had caught the pox from a Gloucester cow named Blossom, who resided in the hamlet of Breadstone. Her hide survives at St George's Hospital medical school in London. Nelmes exhibited "a large pustulous sore" on her hand, offering Jenner a rare source of material with which to test his hypothesis that inoculation with cowpox would offer immunity to smallpox.

BLOSSOM, NELMES, AND PHIPPS

© Wellcome Library, London | Creative Commons

Accordingly, Jenner "selected a healthy boy, about eight years old, for the purpose of the inoculation of the cow-pox." The boy was James Phipps, son of his gardener, and on May 14th, Jenner took "matter" from Nelmes's sore and put it "into the arm of the boy by means of two superficial incisions." After a week Phipps suffered mild fever and aches, but he quickly recovered. Now came the crucial step:

> In order to ascertain whether the boy, after feeling so slight an affection of the system from the cow-pox virus, was secure from contagion of the smallpox, he was inoculated on July 1st following with various matter immediately taken from a [smallpox] pustule... no disease followed. Several months afterwards he was again inoculated with variolous matter, but no sensible effect was produced on the constitution.

IN THE FAMILY
Bucolic depiction of Jenner vaccinating his own infant son, complete with dairy maid and cow peeking through the window.

Jenner had discovered that it was possible to stimulate full immunity to smallpox through inoculation with a harmless disease agent, a process he termed vaccination from the Latin name of cowpox, *vaccinae*, although he had no conception of the true mechanisms. In fact, Phipps's white blood cells had learned to make antibodies to molecules

THE COW POCK
Classic Gilray cartoon satirizing both the new craze for vaccination and popular fears and myths about the process.

VACCINE CLERK TO THE WORLD

on the surface of the cowpox virus, a phenomenon known as acquired or adaptive immunity. When later exposed to smallpox virus sporting the same molecules, Phipps's immune system was able to destroy it with the antibodies, preventing infection.

Jenner was not the first to achieve successful vaccination; Benjamin Jesty, a Dorset farmer, had successfully used the technique to protect his family from an outbreak of smallpox in 1774, and controversy would later arise over precedence between the two men. The dispute is instructive in terms of the scientific meaning of discovery; only Jenner had explored the phenomenon scientifically, proving a cause-and-effect connection rather than a suggestive correlation, and so he is said to be the true discoverer of vaccination.

Although Jenner's initial paper on vaccination to the Royal Society was too revolutionary to be accepted, his privately printed pamphlet of 1798, *An Inquiry into the Causes and Effects of the Variolae Vaccinae, Known By the Name of the Cow Pox*, caused a sensation. By the following year vaccination was in use across Europe and North America; in 1800 it was adopted by the Royal Navy and by 1802 it had reached India. Over the following years Jenner worked tirelessly to spread best practice, publishing four more pamphlets and corresponding with and supplying cowpox material to people across the world. He referred to himself as "Vaccine Clerk to the World." His own practice suffered as a result, and in recognition of his labors the British government awarded him £10,000 in 1802, and a further £20,000 in 1807.

Vaccination against smallpox has been so successful that the last naturally occurring case of smallpox was in Somalia in 1977, and in 1980 the World Health Organization announced: "The world and all its people have won freedom from smallpox." The late 19th century saw Louis Pasteur develop three more vaccines, including one for rabies, and today there are 24 vaccines in worldwide use, saving countless lives.

THE VOLTAIC PILE: THE PRINCIPLE OF THE ELECTRIC BATTERY

1799

Discoverer: Alessandro Volta

Circumstances: Suspecting that Galvani's results with frogs' legs demonstrate not animal electricity but metallic electricity, Volta tries an experiment on his own tongue

Consequences: Invention of the battery, leading to electrolysis

It will be seen that Volta has presented to us a key which promises to lay open some of the most mysterious recesses of nature...

Humphry Davy, first Bakerian Lecture, November 20th, 1806

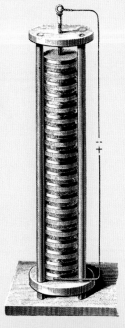

Seeking to disprove Galvani's theories about animal electricity, Alessandro Volta created the voltaic pile, the first battery, a revolutionary piece of technology that would change the world and spark one of the most frenzied bursts of experimentation and discovery in the history of science. He began with a silver spoon in his mouth.

A GENIUS FOR ELECTRICITY

When Galvani sent a copy of his pamphlet "On the Effect of Electricity on Muscular Motion" to the professor of physics at Pavia, Alessandro Volta, he was hoping for the approval of perhaps the leading electrical scientist in the world. Volta was born in 1745, and although his parents had feared that he was a halfwit when he did not learn to talk until the age of four, Volta went on to become something of a prodigy, early conceiving an interest in the mysteries of electricity. Describing himself as having "a genius for electricity," he built his first lightning rod at age 17, and later invented an improved static electricity generator called the "perpetual electrophosphorus," eventually becoming a professor of experimental physics at Pavia in 1778.

A MAN AND HIS PILE
Alessandro Volta shown with the pile of his own devising. Despite being a slow starter (having only learned to speak at age four), Volta went on to become perhaps the preeminent scientist in Italy.

In 1791 he received Galvani's paper, initially dismissing the findings as "miraculous… [and] unbelievable," but colleagues urged him to repeat the experiments and he soon changed his mind. Now he moved, in his own words, "from incredulity to fanaticism"; Galvani's discovery, he said, "proves animal electricity… and places it among the demonstrated truths." But by 1793 he had begun to doubt; further experiments showed that the contractions could be elicited by touching only the nerves with bimetallic contacts, and avoiding the muscles altogether—where did this leave Galvani's Leyden jar analogy? Volta suspected the frog muscle to be acting as a detector of external electrical potential, and he had a strong notion of the true source of this potential. Experimenting on the frog muscle with contacts of just one type of metal failed to produce contractions; only when two different metals were used did the effect materialize. Perhaps it was the contact between two metals that generated the potential, Volta reasoned.

TIN-COATED TONGUES

In fact, this phenomenon, known as contact potential, had already been discovered, though unwittingly. In 1752, German mathematician Johann Georg Sulzer had discovered that pieces of silver and lead, put in contact and placed on the tongue, produced a taste similar to "vitriol of iron" (known as sulfate of iron today). Volta repeated this experiment for himself. According to an early 19th-century science digest, *Memoirs of Science and the Arts*:

> The tongue occurred to Mr. Volta as an organ of voluntary motion, very sensible [i.e. sensitive]... Having covered the point of his tongue... with tin foil, he applied [...] a silver spoon [initially a coin] further on the tongue, and inclined it forward so as to meet the tin. No movement was the result, but he was sensible of a pretty strong sour taste at the tip of his tongue.

He then repeated the trial in a rather gruesome animal model: "Having cut out the tongue near the root of a newly killed lamb, he applied tin foil to the cut part, and a silver spoon to one of its surfaces. On establishing the communication, the tongue was seen to tremble briskly, and its point to rise, fold back, and turn from side to side."

The sour taste indicated to Volta that some form of electrical discharge was taking place; the fact that the taste persisted as long as the metals were on his tongue showed that the discharge was continuous. This was a massive breakthrough; hitherto it had only been possible, using Leyden jars and electrostatic generators, to produce one-off discharges in bursts or sparks. Volta tried wiring different metals to his forehead and ears, producing bright flashes and noises.

IMPERIAL COMMAND PERFORMANCE
Volta demonstrates the workings of his pile to an audience including the Emperor Napoleon, seated at right.

Using an electrometer, a device for detecting strength of electrical potential similar to an electroscope, Volta tested different combinations of metals and found that he could rank them in order of relative charges, with zinc and copper being a particularly good combination. It is now known that copper attracts electrons much more strongly than zinc, so that when the metals are in contact there is a flow of electrons attracted away from the zinc and toward the copper. He also deduced that the potential could be created when the metals were in contact with certain liquids, such as the saliva on his tongue. Such liquids are now called electrolytes.

Volta had now established to his satisfaction that bimetallic contact potential explained Galvani's results, but Galvani and his supporters hit back with experiments supporting their claims. Volta wanted to create a model with no organic component, so he recreated his tongue experiment *in vitro*, as it were, with, in place of his tongue, a disc of cardboard or cloth soaked in salty water. A disc of silver (later copper) and a disc of zinc, separated by a disc of wet material, constitutes a single cell; individually they produce only tiny currents, but the outputs are additive, so a stack or pile of them can produce a significant current. Using vertical glass rods to keep the discs in place, Volta constructed his first pile in 1799. Voltaic piles differed from previous sources of electricity in generating continuous direct current of low potential (or voltage, a unit named for Volta, although he termed it "electrical tension") but high current (or amperage). The wet interstitial discs tended to dry out in a pile, so Volta also contrived a slightly different setup called a *couronne de tasses*, with a ring of cups of brine connected by alternating metal strips of zinc and silver. A series, or battery, of these could be connected together to produce a powerful punch, as would soon be demonstrated.

In March 1800 Volta communicated to the world his discovery of contact potential and the subsequent invention by means of a letter, written in French, to the Royal Society in London. Describing his new *"organe électrique artificiel,"* he argued that animal electricity is a fiction, claiming that even the most famous example, the torpedo or electric ray, is simply a form of voltaic pile: "Not even in this case is it proper to speak of animal electricity, in the sense of being produced or moved by a truly vital or organic action… Rather, it is a simple physical, not physiological, phenomenon—a direct effect of the Electromotive apparatus contained in the fish."

A COUNTRY UNEXPLORED

Almost immediately the revolutionary power of the new technology was demonstrated. Volta had sent his letter in two parts, and before the second half had even arrived two London scientists, Anthony Carlisle and William Nicholson, had built their own pile and discovered that it could "decompose" water into its constituent parts, oxygen and hydrogen. They had discovered electrolysis. The great French chemist Lavoisier had defined an element as a substance "so far undecomposed," recognizing that several substances that so far had been impossible to split into component parts might yet prove to be compounds. The earths potash and soda would, he confidently predicted, prove to be such. Electrolysis offered the potential to show that he had been right.

The man who would most clearly grasp the implications of the new technology was the fiery young British chemist Humphry Davy. In a lecture of 1806, Davy explained that Volta had presented a "key... to lay open [...] nature." Predicting "a boundless prospect of novelty in science; a country unexplored," Davy proposed to use the new voltaic pile as a tool for a new kind of analysis, promising investigations that "can hardly fail to enlighten our philosophical system of the earth; and may possibly place new powers in our reach." He did indeed go on to decompose molten salts with powerful voltaic piles, discovering potassium and sodium by electrolysis of potash and soda, just as Lavoisier had predicted.

Volta himself traveled to Paris in 1801 to present his voltaic pile to a distinguished audience including Napoleon himself. The Emperor assisted in the demonstration, drawing sparks from the pile, melting a steel wire and electrolyzing water. Volta was showered with honors, ennobled, and given a generous pension. In 1819 he retired to an estate in the country. Having previously been such a workaholic that, when engaged in research, the only way that his valet could get him to change his clothes was to distract him with questions about science, Volta now admitted to a friend that he liked domestic life best of all.

CATEGORY

Astronomy

Biology

Chemistry

Exploration

Mathematics

Medicine

Physics

Technology

NO LAUGHING MATTER: THE PAIN-NUMBING EFFECTS OF NITROUS OXIDE

1799

Discoverer: Humphry Davy

Circumstances: A young scientist tests the effects of inhaling various gases—on himself

Consequences: Medical anesthesia and relief from pain for mothers in labor, dental patients, and many others

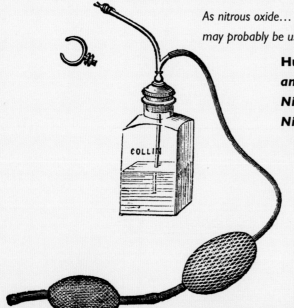

As nitrous oxide... appears capable of destroying physical pain, it may probably be used with advantage during surgical operations...

Humphry Davy, *Researches, Chemical and Philosophical: Chiefly Concerning Nitrous Oxide, or Dephlogisticated Nitrous Air, and Its Respiration* (1800)

© Morphart Creation | Shutterstock.com

COLLIN

One of the most notorious episodes in the Romantic era of science, Humphry Davy's discovery of the anesthetic powers of nitrous oxide, along with its euphoric effects, is a startling and sometimes hilarious tale of reckless self-experiment and philanthropic idealism.

Nitrous oxide (N_2O) is a colorless, almost odorless gas that can trigger pain- and anxiety-relieving mechanisms in the brain, along with pleasurable reward mechanisms, in a similar fashion to opiates, although it is largely nonaddictive. It was probably first made by the Scottish scientist Joseph Black, who wrote of heating ammonium nitrate until it "copiously converted into vapor," which is how the gas is still made in classroom demonstrations today. Black, however, appears not to have recognized what he had made, and it was left to Joseph Priestley in 1772 to discover what he called "diminished nitrous air" by letting "nitrous air" (aka nitric oxide, NO) react with damp iron filings to become N_2O. Priestley could find no use for the new gas, and it joined a long list of new "airs" discovered by him and others toward the end of the 18th century.

This list seemed to offer exciting new horizons in medicine to Dr. Thomas Beddoes, a radical physician in Bristol who in 1798 set up a new philanthropic venture to dispense to the poor and needy the therapeutic benefits of inhaled gases. The Pneumatic Institution for Relieving Diseases by Medical Airs was based on Beddoes's belief that absorption through the lungs enhanced the medical benefits that must surely reside in some of these newly discovered airs, although he was casual in his approach to empirical evaluation. Beddoes employed as Medical Superintendent of the clinic a brilliant, energetic, and ambitious young Cornish scientist, Humphry Davy, who had a more scientific cast of mind, and in April 1799, Davy set about testing various airs for their therapeutic potential. His first experimental subject was always himself, and Davy flirted with death on numerous occasions, such as when inhaling carbon monoxide and finding himself "sinking into annihilation."

Davy settled on nitrous oxide as the safest and most promising of the airs for further research, and set about inhaling increasingly heroic doses while he and his team of assistants monitored his physiological

THE PNEUMATIC INSTITUTE OF DR. BEDDOES

THE ROMANTIC SPIRIT
The classic Thomas Phillips portrait of Davy as a young man, every inch the new Romantic: the scientist with the soul and sensibilities of a poet.

THE AIR IN HEAVEN

and psychological reactions. The gas was an immediate hit, causing euphoria and increased arousal: "Sometimes I manifested my pleasure by stamping or laughing only; at other times, by dancing around the room and vociferating."

Nitrous oxide is still popularly known as "laughing gas," and it quickly became apparent that it offered recreational possibilities. But did it have therapeutic value? Davy increased the dose. On May 7th, 1799, Davy inhaled a massive dose of six quarts (over six liters) of pure gas. "The pleasurable experience… in the middle of the experiment was for a moment so intense and pure as to absorb existence. At this moment, and not before, I lost consciousness; it was however, quickly restored…" He had discovered that nitrous oxide could bring about temporary unconsciousness with no ill effects.

That same month he began trying out the gas on other people, and it was rapturously received by his circle of poetic and learned friends. "I am sure the air in heaven must be this wonder-working gas of delight," the poet Robert Southey famously enthused. "Davy has actually invented a new pleasure, for which language has no name," he wrote to his brother, "I am going for more this evening!" In December Davy inhaled his largest dose yet with the aid of a specially designed gas chamber. After breathing 60 quarts he became "completely intoxicated." After an intense hallucinogenic trip he "stalked majestically out of the laboratory to inform Dr. Kinglake [his assistant] that "Nothing exists, but Thoughts!"

Davy's laboratory notebooks reveal that he had by now grasped the anesthetic qualities of nitrous oxide. The gas could be used for suppressing "intense physical pain," he noted, writing that "Sensible pain is not perceived after the powerful action of nitrous oxide because it produces for the time a momentary condition of other parts of the nerve connected with pleasure." Davy even treated his own toothache, caused by impacted wisdom teeth, with the gas, noting, "The pain diminished after the first four or five inspirations." In 1800 he published the fruits of his investigation as a paper, *Researches, Chemical and Philosophical: Chiefly Concerning Nitrous Oxide, or Dephlogisticated Nitrous Air, and Its Respiration*, which included the prophetic lines: "As nitrous oxide… appears capable of destroying physical pain, it may probably be used with advantage during surgical operations…"

Yet somehow he never made the leap to active pursuit of a practical application that could have changed history, not to mention the declining fortunes of the Pneumatic Institute. In fact, Davy had already lost interest in gas therapy, privately concluding that even nitrous oxide offered no real therapeutic value, and was by now focused on the exciting possibilities offered by the newly discovered voltaic pile (see page 117). He began electrical experiments in Bristol in 1800 and the following year he left the institute to become director of the Chemical Laboratory at the Royal Institution in London.

Davy may have discovered both the analgesic (pain-blocking) and anesthetic (blocking of all feeling/sensation) effects of nitrous oxide, but he had failed to invent medical anesthesia. This great leap would have to wait 40 years.

The recreational effects of laughing gas ensured that it enjoyed continued use, and in the 1840s it was being dispensed to paying audiences in touring shows. One such showman was medical school dropout Gardner Quincy Colton, whose show in Hartford, Connecticut, in 1844 was attended by local dentist Horace Wells. Wells noticed that one of the volunteers called up on stage to inhale the nitrous oxide had smashed his leg when stumbling against a bench, but appeared completely insensible to the pain until the gas wore off. Wells asked Colton to help him with a demonstration, and on the following day Wells had one of his own molars extracted while Colton administered the gas. Wells, on regaining consciousness, said that he had not felt "a pin-prick" and proclaimed "a new era in tooth-pulling."

Unfortunately, Wells's claims were met with suspicion, and a public demonstration at the Harvard Medical School in Boston in 1845 went badly wrong. The patient who had his tooth extracted complained of some discomfort, and Wells was humiliated. His reputation was damaged and he committed suicide three years later. A former colleague

A NEW ERA IN TOOTH-PULLING

© Print Collector | Getty Images

FIGURE OF FUN
Satirical cartoon of a laughing-gas demonstration at the Royal Institution, lampooning "new discoveries in Pneumaticks!" Davy is the young man behind the dais with the thick mop of hair and wild look in his eye.

of his, however, William Thomas Morton, continued to push the case for medical anesthesia, introducing the use of ether, which was more widely available, easier to transport, and less likely to cause death by oxygen starvation. Morton amputated a man's leg under ether at the Massachusetts General Hospital on October 16th, 1846, and a new era dawned in medicine.

Davy's link to the discovery of anesthesia was not forgotten. In December 1846 the first demonstration of surgical anesthesia in Britain took place in Bristol. A surgeon named Lansdowne amputated a man's leg, using ether supplied by the chemist William Herapath. Herapath wrote an account for the Bristol *Mirror*, which suggests he knew of Davy's work: "I have no doubt the inspiration of nitrous oxide [laughing gas] would have a similar effect upon the nerves of sensation as the vapor of ether, as I have noticed that persons under its influence are totally insensible to pain…"

Today nitrous oxide remains the most widely used of all general anesthetic agents, and is used in both dentistry and medicine. It is particularly associated with childbirth as it makes up the "gas" in gas-and-air, a 50/50 mix of nitrous oxide and oxygen, also known as "entonox." According to Dimitris Emmanouil and Raymond Quock, authors of a paper in the journal *Anesthetic Progress*, "The introduction of anesthesia through N_2O is considered a great achievement in dentistry, comparable to the discovery of local anesthesia and the fluoridation of water."

INFRARED LIGHT: ONE EXTREME OF THE VISIBLE SPECTRUM

1800

CATEGORY

Astronomy

Biology

Chemistry

Exploration

Mathematics

Medicine

Physics

Technology

Discoverer: William Herschel

Circumstances: Observing the sun through filters, Herschel realizes that some colors transmit more heat than others

Consequences: Extends the spectrum beyond the visible; unites heat and light

You have undoubtedly heard of Herschel's discovery concerning the production of heat by invisible rays emitted from the sun. By placing one thermometer within the red rays, separated by a prism, and another beyond them, he found the temperature of the outside thermometer raised by more than that of the inside.

Humphry Davy writing to his friend Davies Giddy, July 3rd, 1800

William Herschel's discovery of infrared light is a graphic illustration of both the power of the scientific method to uncover the hidden secrets of nature, and of the role of chance in scientific discovery. In extending the spectrum beyond the visible, Herschel directly inspired the discovery of ultraviolet light by Johann Wilhelm Ritter.

FULL SPECTRUM

If you pass a ray of sunlight through a glass prism, as Isaac Newton had famously done in the 1660s, you will see that it splits apart to give a rainbow of light, with blue light on one side and red light on the other. This is called a spectrum. What you can see, however, is not the entire spectrum; the human eye is only adapted to perceive light in the region of the spectrum in which radiation from the sun is at its peak. For obvious reasons this is known as the visible spectrum. The length, from one peak to the next, of light waves in this part of the spectrum ranges from 31.5–15.8 microinches, roughly the size range of a single cell from your body, or about a 200th the width of a human hair. Light of wavelengths greater than or smaller than this is not visible to the human eye, so that its existence was not suspected until the brilliant enquiring mind of William Herschel, by now the Court Astronomer to George III, was stimulated by a strange finding.

In 1800 Herschel was engaged in the risky task of observing the sun through his telescopes. He had long been fascinated by our nearest star, advancing an eccentric theory "On the Nature and Construction of the Sun and Fixed Stars" in a 1795 paper to the Royal Society, in which he proposed that it was inhabited by intelligent aliens who lived inside its cool interior. This complemented his earlier theories that the moon was inhabited and that the craters on its surfaces were circular housing estates—or circuses—similar to those of Bath, where he had been living. In order to gaze safely at the intense light of the sun, Herschel employed: "… various combinations of differently colored darkening glasses [i.e. filters]. What appeared remarkable was that when I used some of them, I felt a sensation of heat, though I had but little light; while others gave me much light, with scarce any sensation of heat."

Could it be the case that heat was transmitted more by some colors of light, which is to say, parts of the spectrum, than by others? To investigate further he set up an experiment. Like many of the classic experiments from this golden age of scientific discovery, it is wonderfully simple and can easily be repeated by anyone. As Newton had done in his *experimentum crucis* (see page 84), Herschel shone sunlight through a slit in a piece of card to create a single shaft of light, which he passed through a prism—possibly taken from a chandelier)—to create a spectrum. He then placed a thermometer in each color of the spectrum in turn, recording the measurements after eight minutes and finding the average rise for violet, green, and red, respectively, to be 2°F (–16.7°C), 3.2°F (–16°C), and 6.9°F (–13.9°C). Clearly, the red rays of sunlight have a greater heating effect (or there are more of them).

As the sun moved across the sky, so the spectrum cast through Herschel's prism moved across the table, so that the thermometer that had been placed in the red zone was now sitting just beyond the red end of the spectrum (in some versions, Herschel puts it there deliberately). He noticed that the temperature rise was even higher. There must, Herschel concluded, be invisible rays coming from the sun, which could be refracted like the visible colors, and which had a particular power of heating. Further experiments showed that these invisible rays could be reflected as well.

LOOKING FOR LIGHT
William Herschel demonstrating his apparatus for discovering infrared rays; note the huge 40-foot (12m) telescope visible through the window—completed in 1789, it remained the largest in the world for over 50 years.

Today these "invisible rays" are called infrared light, meaning "below-red," because they comprise waves with a frequency lower than that of visible red light. This means they have a longer wavelength, ranging from around 1 micron to about 1 millimeter. This is a broad range, so astronomers, who use infrared light to observe objects too cold to emit higher-frequency shorter-wavelength light, divide the infrared spectrum into subregions: the "near infrared" (39–197 microinches); the "mid infrared" (197–1,181 microinches), the "far infrared" (1,181–11,810 microns), and the "sub millimeter" (11,810 microns to ¹⁄₁₆in), although exact definitions vary.

Herschel recognized that this new form of radiant heat is not a separate category of energy from light; they are simply two forms of the same phenomenon. "We are not allowed, by the rules of philosophizing," he explained, citing the principle of Occam's Razor, "to admit two different causes to explain certain effects, if they may be accounted for by one." Thus, he united heat and light. Radiant heat is simply a form of light;

you can feel it "shining" on you by standing in front of an open oven — the oven appears dark, but you can feel the heat anyway.

In fact, everything in the universe gives off infrared light to some extent (aka glows), because nothing exists at "absolute zero," the temperature of 0 Kelvin that indicates that an atom or body has no energy at all. The human body glows in the mid infrared, while very cold clouds of interstellar gas glow in the far infrared. In 1856 it was discovered by the Astronomer Royal for Scotland, Charles Piazzi Smyth, that the Moon gives off infrared light. In fact, infrared observations of the Moon led astronomers to theorize that it is covered in fine dust — this was in 1948, 20 years before the Moon landings proved the model correct. Piazzi also found that more infrared can be detected by higher altitudes, the first indication that the Earth's atmosphere absorbs much of the infrared coming from outer space, especially very long wavelengths. This is why telescopes sensitive to long wave infrared, such as the Herschel Space Observatory, have to be sent up into space.

The universe itself has a background temperature of about 3 K, and hence gives off a glow in the millimeter range. Even the coldest place in the universe, at 100 picoKelvin, or 0.1 billionths of a K (32°F or 0.0000000001 °C), a minute region of the Low Temperature Laboratory at the Helsinki University of Technology, where the lowest temperature ever was achieved, radiated some infrared.

Herschel applied his thermometer test to the other end of the spectrum, beyond the violet zone, but detected nothing. He did, however, make another prescient suggestion: that different colors of light might affect chemicals in different ways. It was this insight that Johann Wilhelm Ritter used to discover ultraviolet light the following year.

THE AIR UP THERE: JOSEPH GAY-LUSSAC'S HOT AIR BALLOON REVELATION

1804

CATEGORY

Astronomy

Biology

Chemistry

Exploration

Mathematics

Medicine

Physics

Technology

Discoverer: Joseph Gay-Lussac

Circumstances: A balloon flight to test the Earth's magnetic field in relation to altitude

Consequences: Settles question regarding magnetic field; ascertains breathability of high atmosphere; leads to Gay-Lussac's greatest discovery in gases

We observed the animals we had with us at all the different heights, and they did not appear to suffer in any manner. For ourselves, we perceived no effect any more than a quickening of the pulse.

Report of Gay-Lussac and Jean-Baptiste Biot (1804)

The exploits of pioneer balloonists from the 1780s to the early 1800s literally opened a new direction in exploration: up. No one went higher than Joseph Gay-Lussac, who also accomplished the most scientifically worthwhile ascent of the early balloon era.

BALLOON MANIA

When Henry Cavendish discovered hydrogen, he contrived to isolate it by filling soap bubbles with the new gas, immediately noting that it is extremely buoyant, or much lighter than air. Lavoisier was able to measure this property precisely, finding that hydrogen weighed only $\frac{1}{13}$ as much as common air. There was an obvious application for such a gas. Paper-maker Joseph Montgolfier and his brother experimented with hydrogen balloons before settling on hot air, filling a giant paper bag with heated air as a publicity stunt, and drawing huge crowds. They would go on to achieve the first manned flight, on November 21st, 1783, but of greater scientific significance was the ascent, ten days later, of Dr. Alexandre Charles and an assistant in a hydrogen balloon of his own design. Hydrogen balloons have far greater buoyancy than hot air balloons, and so they can carry heavier loads and get higher, faster.

Benjamin Franklin, writing to Joseph Banks in 1783, described the remarkable balloon ascents of the Montgolfiers and Charles, with their rival lifting systems. He speculated that the new technology might "pave the way to some discoveries in Natural Philosophy of which at present we have no conception." For his part, Banks was initially skeptical about what he called "Ballomania," but allowed that if further experiments proved successful, "The immediate Effect it will have upon the Concerns of Mankind [will be] greater than anything since the invention of shipping."

ASCENT OF THE CHARLIER
The inaugural flight of a hydrogen-balloon—the type that became known as a Charlier, after its pilot the scientist Jacques Alexandre Charles, often referred to as just Alexandre to avoid confusion with mathematician Jacques Charles.

Many of the balloon pioneers attempted to put a scientific gloss on their stunts, taking air samples, testing the reactions of animals, and using barometers and other devices to measure altitude and pressure. But in truth there was little immediate utility to the ascents; Samuel Johnson witheringly opined, "I know not that air balloons can possibly be of any use." Perhaps their greatest value lay in the psychological impact of opening a new frontier, a pushing of the envelope that inspired the rising Romantic fervor that would help to drive science in this era.

What is most striking about the early accounts is the sheer wonder and novelty of entering into a new world of space. Benjamin Franklin, the American ambassador in Paris, recorded that, "Someone asked me — what's the use of a balloon? I replied — what's the use of a newborn baby?" Dr. Charles himself related that his first flight brought "a sort of physical ecstasy. My companion Monsieur Robert murmured to me — I'm finished with the Earth. From now on it's the sky for me!"

When he and Robert touched down, Charles recklessly asked his companion to step out of the basket, with the result that the now lightened balloon shot upward at vertiginous speed, ascending to 10,000 feet (3,050 meters) in just ten minutes. "I was the first man ever to see the sun set twice in the same day," observed Dr. Charles, surely a distinction to set alongside the claims of van Leeuwenhoek and his microbes, Galileo and his Jovian moons, or Columbus and the New World.

Born in France five years after the first balloon ascents, Joseph Gay-Lussac was a child of the Revolutionary era, able to take advantage of the new educational opportunities open to bright young men. He went from the new École Polytechnique to the prestigious École Nationale des Ponts et Chaussées (School of Bridges and Highways), and was clearly a star in the making. In 1801 he was poached as an assistant by the eminent chemist Claude-Louis Berthollet, who supposedly told him, "Young man, it is your destiny to make discoveries."

DESTINED TO DISCOVER

Gay-Lussac's primary achievements came in his brilliance at making very precise measurements of the volumes of gases, under different conditions and before and after reactions. In 1804, after Berthollet helped arrange for a hydrogen balloon to be brought over from Egypt, he volunteered to take part in a daring experiment to test suggestive results obtained by previous ascents, relating to whether the strength of the Earth's magnetic field fluctuated with altitude.

BRAVE BALLOONIST
Joseph Gay-Lussac, who enjoyed a brief stint as an intrepid atmospheric explorer, before settling down to a fruitful career in science.

On August 24th, 1804, Gay-Lussac and Jean-Baptiste Biot lifted off from the garden of the Conservatoire des Arts et Metiers, accompanied by a selection of animal copilots. "At 10,000 feet above the ground," their report reads, "we set a little green-finch at liberty. He flew out at once… We followed him with our eyes till he was lost in the clouds. A pigeon, which we set free at the same elevation… launched himself into space… sweeping round and round in great circles, ever reaching lower, until he also was lost in the clouds." As for the aeronauts themselves, who achieved a height of 13,100 feet (4,000 meters), "We were much surprised that we did not suffer from the cold… Our pulses were very quick… Nevertheless, our respiration was in no way interfered with, we experienced no illness, and our situation seemed to us extremely agreeable."

THE MOST ELEVATED POINT OF MY ASCENT

Unfortunately, problems with the flight prevented Gay-Lussac from getting the readings he wanted, so he decided to go up again, this time in a bigger balloon. On September 16th he went up on his own, reaching a height of 23,000 feet (7,016 meters), an altitude record that stood for nearly 50 years. "Having arrived at the most elevated point of my ascent… my respiration was rendered sensibly difficult, but I was far from experiencing any illness of a kind to make me descend. My pulse and my breathing were very quick…" Despite a temperature of minus 49.1°F (9.5°C), Gay-Lussac remained aloft for some time, taking measurements of temperature, humidity, and magnetism.

While a veritable menagerie supposedly accompanied Gay-Lussac and Biot on their flight, including a sheep, cockerel, pigeons, snakes, and bees, Gay-Lussac had rather more prosaic company on his solo flight, in the form of a tatty kitchen chair. At one point, looking to shed ballast, he tossed the chair out of the basket; landing near a village, it occasioned great confusion among the villagers, who assumed it had fallen from Heaven, but could not understand why the angels would employ such scruffy furniture.

Although his magnetic measurements were still less than satisfactory, Gay-Lussac was able to show that the Earth's magnetic field does not vary significantly with altitude. In fact, the strength of a magnetic field declines very quickly over distance, in similar fashion to gravity, but, again as with gravity, because the magnetic field of the Earth is so massive to begin with, this diminution only applies over vast distances. The magnetosphere, the region of influence of the Earth's magnetic field, extends at least ten Earth radii into space—many times more than this in some directions.

Gay-Lussac's record-breaking ascent was more notable for other reasons. First, he had effectively discovered the limit at which life is sustainable and humans can—just about—breathe. More importantly, the subsequent analysis of the air samples he had taken at various altitudes set Gay-Lussac on the path to his greatest discovery, the law of combination by volume, which would prove to be an important confirmation of atomic theory. The law of combining volumes, as it is also known, says that the volumes of gases that react with one another, or are produced in a chemical reaction, are in the ratios of small integers (e.g. 2 volumes of hydrogen + 1 volume of oxygen = 2 volumes steam; 1 volume hydrogen + 1 volume chloride = 1 volume hydrogen chloride). This law was later used by the Italian lawyer and mathematical physicist Amedeo Avogadro to derive the correct formula for water; up until his insights, it had been assumed that water must consist of one hydrogen atom bonded to one oxygen atom, and this in turn had led to incorrect calculations of the atomic weight of oxygen, and many of the other elements in turn. Avogadro showed that the law of combining volumes means that the formula for water must be H_2O, after which it became possible to work out the correct atomic weight for oxygen and many other elements. It also helped to underline the power of John Dalton's atomic theory, which says that each element has its own characteristic atoms, distinguished by their relative weights.

CATEGORY

Astronomy

Biology

Chemistry

Exploration

Mathematics

Medicine

Physics

Technology

THE GREAT FALLS OF THE MISSOURI: LEWIS AND CLARK'S CORPS OF DISCOVERY

1805

Discoverer: The Corps of Discovery

Circumstances: Lewis and Clark battle up the Missouri River in search of great cataracts

Consequences: Discovery of hundreds of new plant and animal species; contact made with many Indian tribes; blazes a trail for America's westward expansion

… to gaze on this sublimely grand specticle [sic]… formes the grandest sight I ever beheld… this truly magnifficent [sic] and sublimely grand object which has from the commencement of time been concealed from the view of civilized man… of its kind I will venture to ascert is second to but one in the known world.

Meriwether Lewis, June 13th, 1805

Perhaps the greatest voyage of river exploration ever undertaken, the Lewis and Clark expedition of 1804–1806 brought back an extraordinary wealth of information, from new animals and plants to geology and ethnography. A daring passage into the unknown, the expedition helped to shape the course of American history.

THE LOUISIANA PURCHASE

In July 1803 President Thomas Jefferson concluded with France one of the greatest land deals in history, the Louisiana Purchase, which saw the young American republic take possession of a huge swathe of territory—nearly 830,000 square miles (2.15 million square kilometers)—stretching from the banks of the Mississippi to the far northwest of North America, almost to the Pacific coast. By this time Jefferson was already planning an ambitious voyage of exploration to extend American claims toward the Pacific and forestall land grabs by European colonial powers. Now the proposed expedition could travel through American-owned land, asserting title and planting the flag as it went, as well as providing a good account of the natural and human resources of the new territories.

In February 1803 Jefferson had petitioned Congress for secret funding for an expedition, and to lead it he had looked close to home, choosing Captain Meriwether Lewis, his private secretary and former neighbor. "It was impossible to find [anyone else] who to a compleat science in botany, natural history, mineralogy & astronomy, joined [together] the firmness of constitution & character... requisite for this undertaking," he later explained. "Beginning at the mouth of the Missouri," Jefferson charged Lewis, "you will take careful observations of latitude & longitude, at all remarkable points on the river, & especially at the mouths of rivers, at rapids, at islands, & other places & objects distinguished by such natural marks & characters of a durable kind." To assist him, Lewis chose an old Army friend, William Clark, who underwent a crash course in natural sciences. Together they would lead a party of some 30 men and, eventually, one woman, in what they called the Corps of Discovery.

In the summer of 1803 Lewis oversaw the construction of a keelboat (a long boat with a shallow draft, suitable for river navigation),

© Photos.com | Thinkstock

MERIWETHER LEWIS
Meriwether Lewis, joint leader of the Corps of Discovery with William Clark. They both went on to careers in politics.

UNKNOWN TERRITORY

and loaded it with provisions and equipment. A major part of his remit was to establish friendly (though colonial) relations with Indian tribes along the way, and to this end the expedition was equipped with a range of "gifts," including 12 dozen pocket mirrors, eight brass kettles, a quantity of vermillion face paint, tomahawks that could double as pipes, and 4,600 sewing needles with accompanying thread. Picking up Clark on the way, the Corps wintered in Illinois and set off in earnest up the Missouri River from St. Charles on May 14th, 1804, poling, dragging, and occasionally sailing against the current.

In early August they met their first Indians, exchanging gifts and establishing good relations, but Lewis and Clark knew it was likely that more hostile tribes lay ahead. At the end of August they reached the Great Plains, marveling at the beauty of the landscape and the diversity of nature. "The Surrounding Plains is open Void of Timber and level to a great extent," wrote Clark in his journal entry for August 25th, 1804. "Numerous herds of buffalow were Seen feeding in various directions; the Plain... extends without interuption as far as Can be seen." In September they endured a tense standoff with Teton Sioux, but escaped unharmed; in fact, the only Corps member to lose his life in the entire expedition was Sergeant Charles Floyd, who had died a month earlier of disease. By winter they had reached a spot near modern-day Bismarck, North Dakota, where they built a log stockade and waited out the cold season.

From here they sent back to Jefferson a cache of specimens, live animals, charts, and detailed reports that alone justified the cost of the expedition. But far more lay ahead as they set off, in the spring of 1805, into completely uncharted territory, never before visited by American explorers or traders.

THE GRANDEST SIGHT

As they penetrated farther up the Missouri, reaching tributaries, they faced hard choices about which path to take. The only way they would know for sure if they had made the correct choice and not strayed from the main river would be to reach the Great Falls of which they had heard tell—a legendary waterfall. In fact, the falls would turn out to be a series of cataracts, but they did not disappoint when Lewis reached them on June 13th, 1805, declaring them "the grandest sight I ever beheld." Shortly after his "ears were saluted with the agreeable sound

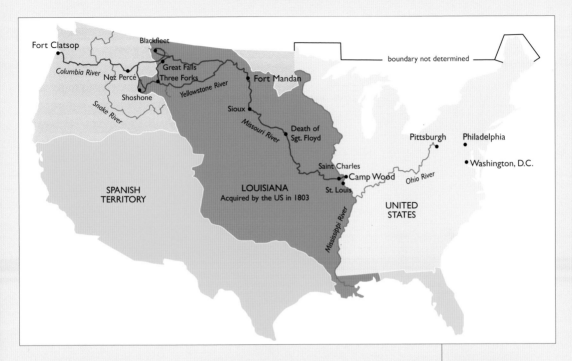

Fort Clatsop

Blackfleet

Great Falls
Three Forks

Columbia River
Nez Percé

Shoshone

Snake River

Yellowstone River

Fort Mandan

Sioux

boundary not determined

Missouri River

Death of
Sgt. Floyd

Pittsburgh   Philadelphia

Washington, D.C.

Saint Charles

Camp Wood   Ohio River

St. Louis

SPANISH
TERRITORY

LOUISIANA
Acquired by the US in 1803

Mississippi River

UNITED
STATES

of a fall of water," he saw "a sheet of the whitest beaten froth for 200 yards in length and 80 feet perpendicular." He described how "The water after descending strikes against the butment... and seems to reverberate and being met by the more impetuous courant they roll and swell into half formed billows of great hight which rise and again appear in an instant."

After carrying their boats around the cascades the party continued upriver, meeting grizzly bears and abundant buffalo, until in August Lewis found the headwaters of the Missouri and soon after saw evidence that he had crossed the Continental Divide, the watershed dividing rivers that run east to the Atlantic from those that run west to the Pacific. "An Indian called me in to his bower and gave me a piece of a fresh salmon roasted," he related. "This was the first salmon I had seen and perfectly convinced me that we were on the waters of the Pacific Ocean." One of the aims of the mission was to blaze a trail for riverborne trade between the coasts, although Lewis was to be frustrated at the lack of a connecting waterway between the Missouri and its Pacific-reaching counterpart, the Colorado River.

THE JOURNEY OF THE CORPS OF DISCOVERY
The path of the Lewis and Clark expedition, following the course of the Missouri to the Continental Divide, then onto the Pacific coast and back. Note the comparative sizes of the Louisiana Purchase and the pre-Purchase United States.

GREAT FALLS
The Great Falls of the
Missouri River, described by
Lewis as a "truly magnificent
and sublimely grand object."

Nonetheless, after battling across the Rocky Mountains and enduring near starvation, the Corps did win through to the Colorado and hurtled downstream to reach the Pacific on November 20th, 1805. "Ocean in View! O the joy," rhapsodized Clark, although in fact he was referring to a misleading sighting of the Colorado estuary. After wintering on the Pacific coast, the Corps struggled back across the Rockies. They made much better time on their return leg, despite detours to explore the Yellowstone and Marias Rivers, arriving back in St. Louis on September 23rd. After firing off their cannon "as a Salute to the Town," recorded Clark, "we were met by all the village and received a harty welcom from its inhabitants &c." Having been given up for dead, the expedition was given a heroes' welcome.

On their epic voyage the Corps had discovered 122 animal species and subspecies and 178 plant species, all previously unknown to science. Lewis enumerated some of the larger mammals they had encountered:

> … the grizzly bear… the black bear, the common red deer, the black-tailed fallow deer, the Mule deer, Elk, the large brown wolf, the small wolf of the plains, the large wolf of the plains, the tiger cat, the common red fox, black fox or fisher, silver fox, large red fox of the plains, small fox of the plains or kit fox, Antelope, sheep, beaver, common otter, sea Otter, mink… seal, racoon, large gray squirrel, small brown squirrel, small gray squirrel, ground squirrel, swelel [mountain beaver], Braro [badger], rat, mouse, mole, Panther, hare, rabbit, and polecat or skunk.

They had established relations with many Indian tribes, revolutionized the study of American natural history, and profoundly shaped the national discourse on the western frontier. The decades to come would see a headlong expansion of American settlement and dispossession of the native tribes. One of the original drivers of this wave of exploitation was the careful and detailed analysis of natural resources provided by Lewis and Clark, whose keen gaze assessed the lands through which they traveled for soil quality, abundance of game, likely mineral deposits, and native populations.

Meriwether Lewis became Governor of the Louisana Territories in 1807, but killed himself in 1809 in the grip of depression. Clark ran unsuccessfully to be governor of Missouri in 1821.

ELECTROMAGNETISM: WHY ELECTRIC CURRENT DEFLECTS A COMPASS NEEDLE

1820

CATEGORY

Astronomy

Biology

Chemistry

Exploration

Mathematics

Medicine

Physics

Technology

Discoverer: Hans Oersted

Circumstances: Convinced that there must be a connection between electricity and magnetism, Oersted decides, in the middle of a lecture, to see if he is right

Consequences: Unites two fundamental forces, inaugurating the electrical age

One feels that magnetic forces are as general as electric forces.
An attempt should be made to see if electricity… has an
action on the magnet…

Hans Oersted, *Ansicht der*
chemischen Naturgesetze (1812)

One of the most remarkable discoveries in the history of physics was made during a lecture, before a hall full of students, yet it passed almost unremarked and appeared so unimpressive that even its discoverer did not follow it up for three months. Yet Oersted's finding that an electrical current can affect a magnet proved to be the starting point of the electrical era.

OF MAGNETS   Magnetism is as familiar as a fridge magnet or a compass needle, yet it is also a force that it is still not fully understood by physicists, and which is difficult to explain in any detail. The phenomenon of magnetism, whereby some materials have a power of attraction or repulsion to other magnets and to some metals, has been known since ancient times; the word "magnet" comes from the Greek region Magnesia, where lodestone was mined in antiquity. Lodestone is an iron ore with naturally occurring magnetic properties. Magnetism became essential to global trade, exploration, and shipping with the invention of the compass (by at least the 11th century CE). Lodestone could be used to make compass needles; another direct confirmation of the link between compasses and magnetism came in 1269, when Pierre de Maricourt made a globe of lodestone and showed that a compass needle aligned itself according to mysterious lines of influence around the globe.

THE PHARMACIST
The son of a pharmacist, Hans Oersted trained in the same profession. A scholarship enabled him to travel round Europe in the early 1800s, exposing him to the latest discoveries in the field of electricity.

This experiment was extended in a study by English physician William Gilbert, described in his landmark 1600 book *De Magnete*. In it, Gilbert described how he constructed a globe of lodestone called a *terella* ("little earth"), and followed its influence on a *versorium*, a freely pivoted metal needle. Noting that the dipping, or inclination, of the needle as it moved around the globe marked out lines of influence that ran "north–south" like lines of longitude, Gilbert reasoned by analogy that a similar mechanism explains the inclination of compass needles as they move about the globe, and that the Earth must therefore be a *Magno Magnete*, or "great magnet."

Gilbert was also responsible for coining the term electricity, from the Greek word for amber, which was known to develop an electric charge through friction. Importantly, and in opposition to contemporary assumptions, he demonstrated that although both static electrical charges and magnetism have powers of attraction and repulsion, they are different kinds of force. At the same time, however, there was evidence to suggest a link between magnetism and electricity. It was known that when lightning struck a ship, the polarity of the vessel's compass might be reversed. Scientists in the 18th century were particularly aware of an incident involving a cutlery tradesman from Wakefield in England. The hut in which he was staying was struck by lightning, and the knives and forks that survived the strike became magnetized so that small nails were attracted to them. The experiments of Benjamin Franklin suggested that lightning was a form of electrical current, and in 1751 he went further by showing that sewing needles could be magnetized by means of a discharge from a Leyden jar, analogous with a lightning strike.

Thus, by the late 18th century it was widely suspected that there might be a link between magnetism and electricity, but attempts to show one by the obvious method of placing a compass needle near an electric wire had failed, partly for a devastatingly simple reason. The prevailing scientific worldview was Newtonian, following Newton's description of gravity as what was known as a central force: a force acting in straight lines between points, typically radiating out from a central point in straight lines. Work by Joseph Priestley in 1767 suggested that the forces of attraction/repulsion in static electricity did indeed operate along Newtonian lines, following the inverse square law. Accordingly, it was assumed that a magnetic field produced by an electric current flowing along a straight wire would be in the same direction as the flow of current. Hence the way to detect the effect of such a field would be to place a compass needle at right angles to the wire, so that the needle would be deflected by the magnetic action of the current and turn to become parallel to the wire. Since, as would shortly be shown, the magnetic field in such a case actually lies at right angles to the wire, the needles in these experiments were already aligned with them and would not show any effect. Combined with the weakness of electrical currents available to early experimenters, this simple but fundamental flaw meant that the relationship between magnetism and electricity went undetected.

AN ATTEMPT SHOULD BE MADE

Not everyone was blinkered by these Newtonian preconceptions. Danish scientist Hans Christian Ørsted (usually written Oersted in

© UniversalImagesGroup | Getty Images

ELECTROMAGNETISM DEMONSTRATED
Oersted demonstrates to students the effects on a freely rotating needle of a current passing through the wire he is holding.

English) was a follower of Immanuel Kant, having picked up a Romantic philosophy of science while visiting Germany, where he came under the influence of Johann Wilhelm Ritter. Kant argued for unity of natural laws, and following this line of philosophy, Oersted believed that all different forces are simply manifestations of fundamental attractive and repulsive forces. More specifically, he argued in 1812, "One feels that magnetic forces are as general as electric forces. An attempt should be made to see if electricity... has an action on the magnet..."

Oersted was aware of the effect of lightning on compass needles, and of Benjamin Franklin's work suggesting that lightning was a form of electric current. Accordingly, he conceived of a wire carrying a current as acting like a lightning bolt (on the basis that such a wire will glow when current is passing through it). In an article for the *Edinburgh Encyclopaedia* of 1827, in which he discussed the background to his own work in the third person, Oersted explained: "... if it were possible to produce any magnetical effect by electricity, this could not be in the direction of the [axis of the] current, since this had been so often tried in vain, but that it must be produced by a lateral action."

Oersted believed electric current to be an "electric conflict" between decompositions and recompositions of positive and negative electricities. Such conflicts would not be restricted to the conductor itself (i.e. the wire), but would also act in the space around it, forming vortices or zones of circular effect around the wire. This was an early form of field theory, which led Oersted to predict that he would be able to detect an effect on a compass needle where others had failed.

Setting up his materials and equipment for a lecture in April 1820, Oersted provided himself with a compass needle, a voltaic battery (in trough form), and lengths of wire, although he did not intend to try the experiment then and there. But as he gave his lecture he changed his mind and tried the demonstration. As current flowed through a wire Oersted moved the compass close to it, and he and the audience saw that the needle was very slightly deflected to the east above the wire, and to the west below it.

There were no exclamations of eureka, however. Neither audience nor lecturer were overly impressed by the flicker of motion evidenced by the compass needle. "The effect was certainly unmistakable," he later explained, "but still it seemed to me so confused that I postponed further investigation to a time when I hoped to have more leisure." In fact, it was to be three months before Oersted returned to the topic, now armed with a more powerful voltaic battery, and saw the effect much more clearly. He quickly wrote up his findings in a landmark paper, *"Experimenta circa effectum conflictus electrici in acum magneticam"* ("Experiments on the effect of an electric conflict on the magnetic needle"), in which he explained his theory about the circular magnetic field created by an electric current:

> ... we may collect that this conflict performs circles; otherwise one could not understand how the same portion of the wire drives the magnetic pole toward the East when placed above it and drives it toward the West when placed under it. For it is the nature of a circle that the motions in opposite parts should have an opposite direction.

Oersted's paper was widely disseminated, and although French chemist Pierre Louis Dulong described how Oersted's report "is received very coolly here," dismissing it as "a German dream," its effect was electric. At a meeting of the Académie des Sciences in Paris in September, the distinguished scientific statesman François Arago discussed the discovery. In the audience was the French mathematician and scientist, André-Marie Ampère, who replicated Oersted's experiment within a week and developed a mathematical theory describing how the magnetic field depends on the strength of the current and the distance from the wire. The Scottish mathematical physicist James Clerk Maxwell would later describe Ampère as "the Newton of electricity." With a rigorous mathematical description of the relationship between electricity and magnetism, the two forces had been united into one: electromagnetism.

THE NEWTON OF ELECTRICITY

CATEGORY

Astronomy

Biology

Chemistry

Exploration

Mathematics

Medicine

Physics

Technology

THE FIRST DINOSAUR FOSSILS: MRS. MANTELL UNEARTHS IGUANODON TEETH

1820

Discoverer: Mary Ann Mantell

Circumstances: The wife of avid fossil collector Gideon Mantell notices some strange-looking stones amid the gravel of a freshly laid road

Consequences: The discovery of dinosaurs and a world of extinct monsters

The [fossils] are from the lower extremities of the creature and being such a magnificent group,
I have no doubt they are of an Iguanodon. With such excitement I confess I went to bed very late.

Gideon Mantell

© Dorling Kindersley | Thinkstock

Monsters once walked the earth, yet until the turn of the 19th century, they were known only in myths and legends. From around 1770, however, geologists and zoologists began to look at curious rock formations through the new lens of science, realizing that here lay concrete evidence of a long past era of terrible beasts and colossal monsters. In 1820 a chance finding by the wife of the foremost fossil hunter of the age uncovered some of the first vestiges of a group now familiar to every schoolchild: the dinosaurs.

Fossils had been known since antiquity, but could only be interpreted according to prevailing worldviews. Hence they were assumed to be the bones of giants, passingly mentioned in the Bible, the remains of dragons, or evidence of the great flood common to many world religions and cultures. The advance of geology as a science, with increasing realization that the landscape is the result of aeons of change and upheaval, and that rock strata reveals chronology, changed the way in which fossils were understood.

SEA MONSTERS

Among the first discoveries that brought the realization of a lost world of terrifying monsters was that of a huge aquatic reptile, now known as a mosasaur, meaning "Meuse lizard," named for the region in which it was found: the valley of the Meuse River. In the gallery of a subterranean quarry at St. Pietersburg near Maastricht, workmen uncovered a 6-foot (1.8-meter) long skull. In his 1799 *Natural History of St. Peter's Mount*, Faujas de Saint-Fond gives an account of the find:

SEA MONSTER
The dramatic discovery by quarrymen at Maastricht of the colossal skull of a prehistoric sea monster, later named the mosasaur.

> … at the distance of about 500 paces from the principal entrance, and at 90 feet below the surface, the quarrymen exposed part of the skull of a large animal imbedded in the stone. They stopped their labors to give notice to Doctor Hoffman, a surgeon at Maastricht, who had for some years been collecting fossils from the quarries… Doctor Hoffman, observing the specimen to be the most important that had yet been discovered, took every precaution to secure it entire.

The specimen went on to have a chequered history; it promptly became the subject of a lawsuit, before coming to the notice of an acquisitive officer during the Napoleonic Wars. "The [owner], suspecting the

object of this attention, had the skull removed and concealed in a place of safety in the city. After the French took possession of the city… a reward of 600 bottles of wine was offered for its discovery. The promise had its effect, for the next day a dozen grenadiers carried off the specimen in triumph… and it was subsequently conveyed to the museum of Paris." In fact, this was the not the first mosasaur fossil found in the quarry, but the impressive skull would become the type specimen (the fossil first used to describe and name the species), and the creature later acquired the Latin name *Mosasaurus hoffmanni*. The great anatomist Georges Cuvier, the world's foremost expert on fossils in the late 18th century, recognized it as the remains of a huge reptile, and other experts pointed to its similarity to existing monitor lizards. Today, the mosasaur is indeed thought to have evolved from a monitor lizard-like ancestor.

Mosasaurs were large marine reptiles, sometimes known as the "T rex of the sea," which flourished from around 85–65 million years ago, before being wiped out in the extinction event that marked the end of the Cretaceous Era. But they were not dinosaurs. These famous land animals were not discovered for another 50 years.

LEGENDS OF THE FIND

In the early 1800s the foremost fossil collector in Britain, if not the world, was a country doctor from Lewes in East Sussex, Gideon Mantell. Also a keen student of geology and paleontology (the study of prehistoric life), Mantell collected fossils himself and via a range of contacts, who would bring him specimens from quarries. In 1815 a similar arrangement had brought into the hands of William Buckland, Reader in Geology at Oxford University, the first of a series of fossil bones of an unknown large animal. By 1824 Buckland had acquired enough of the bones to announce his discovery of a great prehistoric reptile, which he called megalosaurus ("great lizard"). It was the first dinosaur to be identified, and although it was impressively large and fierce looking, with the bulk of an elephant and sharp teeth in its large jaw, it did not appear so very different from large living reptiles, such as the crocodile or monitor lizard.

More impressive would be evidence of a large, extinct herbivorous lizard, quite unlike anything known to exist in the present, and this is what would bring fame and misfortune to Gideon Mantell. In 1820 he acquired the teeth of a large unknown herbivorous animal. According to scientific legend, created by Mantell himself, it was his wife Mary Ann who first found the teeth, although he told several different versions of the story. The most popular variant sets the finding in 1822, and has Mrs. Mantell amusing herself by taking a walk while her husband visits a patient in the country. Poking at a bush, or possibly scanning the freshly turned rocks of a newly laid road, depending on the version, Mrs. Mantell noticed some fossil teeth, which she handed to her husband.

According to Mantell's biographer, Dennis Dean, whether or not this tale is true it must postdate the actual first discovery of the teeth, which were given to Dr. Mantell by a Mr. Leney in 1820. In his journal Mantell records that on February 25th, 1820, "Mr. Leney brought me a packet of fossils from Cuckfield." The entry for August 9th records: "Received a packet of fossils from Cuckfield." Cuckfield was the site of a stone quarry, and Mr. Leney was possibly the foreman there, or an old family friend who was acting as intermediary between Mantell and one of the quarry foremen who knew that any promising finds would be remunerated by the fossil-mad collector. If Mary Ann did find some teeth, it was probably during a trip to Cuckfield made by the Mantells in August 1820.

Mantell identified the teeth as belonging to an unknown large herbivorous reptile. Over the next two years Mantell collected further bones and teeth of the animal, and in 1822 he was excited to announce what he must have recognized as a potentially revolutionary find to a meeting of the Geological Society. His reception was devastating, with his teeth dismissed as probably those of a mammal, and not that old either. Mantell sent some of the teeth to Cuvier in Paris, but the great man's initial reaction was no more encouraging: he thought they belonged to a prehistoric rhinoceros. Mantell persisted, however, and succeeded in changing Cuvier's mind. He later proudly recalled Cuvier telling him: "These teeth are certainly unknown to me; they are not

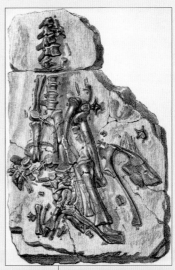

THE MANTELL PIECE
The celebrated "Mantell piece," aka the Maidstone iguanodon: an impressive collection of fossil bones gathered in a single slab, dug up from a quarry in Maidstone.

WE HAVE HERE A NEW ANIMAL

from a carnivorous animal... Do you not agree that it looks as if we have here a new animal, an herbivorous reptile?"

Taking his teeth to the Royal College of Surgeons where a curator noted that they looked much like the teeth of the iguana, Mantell came up with the name iguanodon (literally "iguana tooth"), and in 1825, he published *Notice on the Iguanodon, a Newly Discovered Fossil Reptile, from the Sandstone of Tilgate Forest, in Sussex*, which caused a sensation and made him famous. Over the years he acquired more iguanodon bones; in 1834 a particularly fine collection of bones was found at a quarry in Maidstone. Mantell reported: "They are from the lower extremities of the creature and being such a magnificent group, I have no doubt they are of an Iguanodon. With such excitement I confess I went to bed very late." The slab of rock in question was dug out in one piece and took pride of place on his wall, where it was known as the "Mantell piece."

In 1842 the superintendent of the Natural History Museum, Richard Owen, announced that both megalosaurus and iguanodon were members of a new "distinct tribe or suborder of Saurian Reptiles, for which I propose the name of Dinosauria." Owen now tried to take all the credit for the discovery of the dinosaurs, including the iguanodon, engendering a bitter dispute with Mantell, who had by now been divorced by his wife and crippled by a serious illness compounded by a horrific traffic accident. Mantell condemned Owen's behavior as "unworthy piracy... what a pity a man of so much talent should be so dastardly and envious." Racked with pain both physical and emotional, Mantell died in 1852 of a massive overdose of opium, which might well have been suicide.

NATURAL SELECTION: DARWIN'S THEORY OF EVOLUTION

1835

CATEGORY

Astronomy

Biology

Chemistry

Exploration

Mathematics

Medicine

Physics

Technology

Discoverer: Charles Darwin

Circumstances: Darwin notes the proliferation of small bird species in the Galapogos

Consequences: Develops the theory of evolution by natural selection, revolutionizing science and conceptions of humanity's place in the universe

Seeing this gradation and diversity of structure in one small, intimately related group of birds, one might really fancy that from an original paucity of birds in this archipelago, one species had been taken and modified for different ends.

Charles Darwin, *Journal of Researches*, 2nd edition (1845)

There was no single eureka moment for Charles Darwin's discovery of the theory of evolution by natural selection, but there are two pivotal milestones in his long journey toward the theory. The first came during a literal voyage of discovery, when Darwin was collecting bird specimens in the Galápagos in 1835, and the second came while he was reading a book in London in 1838.

THERE NEVER WAS A FINER CHANCE

"Capt. F wants a man," began a letter to Charles Darwin from his friend and former teacher, John Stevens Henslow, in August 1831. Henslow had been offered a position as a naturalist on a round-the-world voyage of surveying and exploration under Captain Robert FitzRoy, but recommended instead that his former student would be the ideal man. "The Voyage is to last 2 yrs. & if you take plenty of Books with you, any thing you please may be done. You will have ample opportunities at command. In short I suppose there never was a finer chance for a man of zeal & spirit," continued Henslow. Darwin seized the opportunity, and on December 27th, 1831, he sailed out of Plymouth aboard the *Beagle*, on a historic voyage that would take him to Atlantic islands, around South America, across the Pacific via the Galápagos, and home via South Africa.

ICON OF EVOLUTION
The iconic beetle-browed profile of Charles Darwin in a photo probably dating to 1874. Despite an avowed distaste for the tedium of sitting for a portrait, Darwin was repeatedly photographed.

The voyage of the *Beagle* would actually take five years, not two, and for Darwin, who suffered horribly from seasickness, the ship was akin to a floating prison. Although he got on well with FitzRoy he never missed an opportunity to spend time on land, and in fact he would spend just 18 months of the five-year trip aboard ship. Much of this time on land was spent exploring South America, from the lush tropical jungles of Bahia, now Salvador, to the grassy pampas, where he lived the *gaucho* (cowboy) life, to the wild terrain of Tierra del Fuego at the southern tip of the continent. Darwin amassed a vast collection of rock, plant, and animal specimens, and became an avid fossil collector; by the time he returned home his collection included 5,436 skins, bones, and carcasses. He also made detailed investigations of geology, botany, zoology, ecology, paleontology, and anthropology, filling a 770-page diary and 1,750 pages of notes. Along the way Darwin found clues that would set him on the path to his famous theory.

For instance, in August 1833 while fossil hunting at Punta Alta (in what is now Argentina), a spot where he had previously found the bones of extinct large mammals, Darwin uncovered the complete skeleton of an unknown creature (later identified as a giant ground sloth). It was embedded in the cliff face below a layer of white sea shells, indicating that what had been dry land when the animal was alive had subsequently become seabed, and was now once more dry land. How long ago must the creature have lived? Why were there no animals like it in South America in the present day? Evidently, it was an extinct species—why had it died out?

LIVING LAB
The stark landscape of the Galápagos Islands, a natural laboratory for speciation and evolution.

Nearly two years later, in what is now Chile, Darwin experienced a massive earthquake, which struck Valdivia on February 20th, 1835. Exploring the devastated coastline, he observed that it had risen by a few feet, and that higher up from the shore was evidence of prior rises. Here was graphic evidence of what Darwin had been reading about, the work of geologist Charles Lyell, who contended that mountain ranges and cliffs rose up in increments. Earlier in his voyage Darwin had seen fossilized trees embedded in sandstone, which he knew had formed from seashore mud, but at an altitude of 7,000 feet (2,100 meters). It must have taken aeons for the mountains to have risen this high; now he was equipped with the concept of "deep time"—the many millions of years necessary for evolution to produce changes in tiny increments.

On September 7th, 1835, *HMS Beagle* set out from Callao, Peru, to sail to the Galápagos Islands. In August Darwin had written to his sister, Catherine, "I am very anxious to see the Galápagos Islands—I think both the Geology & Zoology cannot fail to be very interesting." He was not disappointed, noting, "The natural history of this archipelago is very remarkable: it seems to be a little world within itself; the greater number of its inhabitants, both vegetable and animal, being found nowhere else." Shuttling between the islands Darwin collected specimens. "I shall be very curious to know whether the Flora belongs to America, or is particular," he wrote to Henslow. "I paid also much attention to the Birds, which I suspect are very curious."

A VERY REMARKABLE ARCHIPELAGO

The birds that attracted his attention most were the mockingbirds; he was struck by the fact that there were different types on each of the four islands. Contrary to scientific legend he did not immediately recognize the proliferation of finch species, not realizing that the birds he variously labeled wrens, "gross-beaks," finches, and oriole-relatives were all finches, and he failed to label his specimens by island. Yet, without properly realizing it at the time, he was picking up clues to the extraordinary speciation at work in the Galápagos: local informants told him that each island had unique types of tortoise and plants.

"IT AT ONCE STRUCK ME THAT UNDER THESE CIRCUMSTANCES FAVORABLE VARIATIONS WOULD TEND TO BE PRESERVED & UNFAVORABLE ONES TO BE DESTROYED"

It was on Darwin's return to England that the pieces fell into place, regarding the Galápagos finches. He had each aspect of his massive haul of specimens analyzed and classified by experts, and the ornithologist John Gould revealed that Darwin's many different small birds were in fact all species of finch: 14 different species. Fortunately, Captain FitzRoy and Darwin's assistant Syms Covington had kept better records than he had, and he was able to reconstruct their distribution. "Seeing this gradation and diversity of structure in one small, intimately related group of birds," he noted in his book *The Voyage of the Beagle*, originally titled simply *Journal of Researches*, "one might really fancy that from an original paucity of birds in this archipelago, one species had been taken and modified for different ends." To the pious FitzRoy, this appeared "to be one of those admirable provisions of Infinite Wisdom by which each created thing is adapted to the place for which it was intended." To Darwin,

INTO THE WEST
The *Beagle* in the Straits of Magellan. Darwin did not care for life onboard and spent remarkably little of his five-year trip on the ship.

however, the finches raised a host of questions. Why were there so many different species? Why were they so similar to species on the mainland, yet different enough to constitute entirely separate species? The prevailing theory was that God had created species as fixed entities, with limited variation within the species. Darwin's grandfather Erasmus had run into trouble for suggesting that species could transmute (we would say evolve) into other species. The Galápagos finches seemed to suggest that this was exactly what had happened. The most logical explanation was that the archipelago had been colonized by a finch species from the mainland, and that the subtle differences in ecology between the different islands had led to what is now called adaptive radiation, where a single species evolves into many different species to exploit all the niches in an ecosystem. Today, the Galápagos is recognized as a "living laboratory" for speciation, and the finches are considered to be the world's fastest-evolving vertebrates, acquiring remarkable diversity of appearance and behavior in just a few million years.

Darwin's finches were powerful evidence for the evolution of new species, but what was the mechanism of this transmutation? The moment that came closest to a eureka for Darwin was reading an essay by the economist Thomas Malthus, "On the Principle of Population," in September 1838. Discussing human population, Malthus contended that while population rises geometrically, food production only rises arithmetically, so that a reckoning of famine and death was inevitable.

BEING WELL PREPARED

In a flash, Darwin saw that in such a "struggle for existence," only the fittest would survive. At the end of his life he recalled, "being well prepared to appreciate the struggle for existence… it at once struck me that under these circumstances favorable variations would tend to be preserved & unfavorable ones to be destroyed. The result of this would be the formation of new species." Darwin called this modified Malthusian mechanism "natural selection," for by it nature acted like the breeder in animal husbandry, selecting the traits best adapted for success, except that in natural selection the process was blind, operating over vast periods of time on large populations in which "the surviving one of ten thousand trials" would pass on its improvement, to effect evolution in tiny increments.

Darwin's theory of evolution through natural selection would change not just biology but all of science, as well as wider society and culture. Since 1858, when he finally published his theory, Darwinism has gone from strength to strength, and been applied to and abused for many ends. Ideas and theories derived from Darwinism, often wrongly, have been used to justify eugenics programs and colonial exploitation. For instance, during their late-19th-century carve-up of Africa, European colonial powers justified their actions with pseudoscientific claims about the evolutionary superiority of the white man.

CATEGORY

Astronomy

Biology

Chemistry

Exploration

Mathematics

Medicine

Physics

Technology

THE AGENT OF INFECTION: JOHN SNOW IDENTIFIES THE CAUSE OF CHOLERA

1854

Discoverer: John Snow

Circumstances: When a violent outbreak of cholera hits London, a leading doctor takes the chance to put to the test his radical new theory about the transmission of the disease

Consequences: Advances in public health; the founding of the science of epidemiology

As soon as I became acquainted with the situation and extent of this irruption [sic] of cholera, I suspected some contamination of the water of the much-frequented street-pump in Broad Street.

John Snow, 1855

John Snow and his map of cholera victims have become medical legends. In the popular version Snow heroically identifies the source of infection and intervenes, single-handedly stopping a cholera epidemic in its tracks. Although the reality is less detective fiction and more science fact, it is true that Snow is regarded as the father of epidemiology and a hero in the field of public health.

John Snow was born in York in the north of England in 1813. As a young man he was apprenticed to a surgeon-apothecary in Newcastle and there witnessed the ravages of the first cholera epidemic to hit Britain, in 1833. By the time the second great cholera epidemic arrived in the late 1840s, Snow was a physician in London, where he had made a name for himself in the new field of anesthesia. Snow had quickly realized the potential of the new ether technology first trialled in the US (see page 126), and had devised improvements in the procedure for delivering ether vapor that helped to make it more popular. He became the UK's leading expert on anesthesia and received the ultimate imprimatur when asked to administer chloroform to Queen Victoria at the birth of Prince Leopold in 1853.

Cholera is a highly contagious, extremely dangerous disease that can strike—and kill—with terrifying speed. It is now known that the disease organism, a bacterium called *Vibrio cholerae*, produces toxins that cause the lining of the gut to pump out fluid, leading to massive dysentery that can be fatal in hours. In the mid-19th century, however, the cause was universally believed to be noxious airs or "miasmas," probably caused by rotting sewage. Cholera can be treated very effectively with simple oral rehydration, but this was not known in the 1800s, and the death rate could be brutal. The epidemic of the late 1840s carried off 14,600 people in London, with a death rate of 6.2 per 1,000 Londoners. It was particularly severe in the Berwick Street area of Soho, where Snow lived, and in 1849 he published a paper on the disease and a monograph, *On the Mode of Communication of Cholera*.

Radically dissenting from the orthodox view of cholera transmission, Snow argued on clinical grounds that the disease must be waterborne, not airborne, since it affected the gastrointestinal system, not the

BY ROYAL APPOINTMENT

© Rsabbatini | Creative Commons

MAN OF THE NORTH
John Snow was a doctor from the north of England whose first-hand experience of previous cholera epidemics drove his desire to discover the cause of the disease, making prevention possible.

respiratory system. He suspected that it was being transmitted by water contaminated with some *"materies morbi."* But, he pointed out in his review of literature on cholera epidemics, "as we are never informed in works on cholera what water the people drink I have scarcely been able to collect any information on this point." Reviewers dismissed Snow's book. "There is, in our view, an entire failure of proof that the occurrence of any one case could be clearly and unambiguously assigned to water," said the *London Medical Gazette*. In 1853–54 Snow would get the chance to put his theory to the test.

THE GRAND EXPERIMENT AND THE GUARDIANS

In 1853 cholera returned to the capital and Snow tested his hypothesis on its mode of transmission by undertaking a comparative study of mortality rates in two parts of south London. One area was served by a water company that drew its water directly from sewage-contaminated parts of the Thames, while another area received water from a company that sourced it from higher up the river, where the water was purer. Snow personally conducted the door-to-door surveys necessary for ascertaining the water source involved in each case of cholera, in what became known as his Grand Experiment. As he suspected, his results showed that customers of the "dirty water" company were eight to nine times more likely than the "upriver" company customers to have died of cholera.

This was solid work in what would become recognized as a new field of public health and science in general: epidemiology, the study of the patterns and causes of disease and health. More dramatic developments now followed. On the night of August 30–31st, 1854, an outbreak of cholera exploded in a small region of Soho, around Broad Street and Golden Square. Overnight, 200 people died in the fastest and deadliest epidemic ever seen in Britain, including whole families struck down at once. The epidemic continued for ten days, eventually killing 616 people, almost all of them from within an area of just a few streets. Snow called it "the most terrible outbreak of cholera which ever occurred in this kingdom." Immediately, he "suspected some contamination of the water of the much-frequented street-pump in Broad Street."

The cause of the outbreak, it would later be discovered by the Reverend Henry Whitehead as he sought to disprove Snow's theory, was a soiled diaper. Whitehead interviewed a woman who lived at 40 Broad Street, whose baby had tragically died of cholera, contracted from elsewhere at the end of August, and who had washed the baby's soiled linens in water that she then emptied into a drain that fed into a leaky cesspool just 3 feet (1 meter) from the Broad Street pump. Whitehead would become one of the most fervent champions of Snow's work.

In September 1854 Snow knew none of this, yet he already suspected contamination of the pump to be the cause. He inspected water from the pump and spoke to a local who told him it had smelled foul a day earlier. On September 5th he obtained from the General Register Office a list of the 83 cholera deaths in the Golden Square area since August 31st, and turned detective:

"WATER! WATER! EVERYWHERE; AND NOT A DROP TO DRINK."

© Print Collector | Getty Images

> On proceeding to the spot, I found that nearly all the deaths had taken place within a short distance of the [Broad Street] pump. There were only ten deaths in houses situated decidedly nearer to another street-pump. In five of these cases the families of the deceased persons informed me that they always went to the pump in Broad Street, as they preferred the water to that of the pumps which were nearer. In three other cases, the deceased were children who went to school near the pump in Broad Street…

The evidence was clear: "there has been no particular outbreak or prevalence of cholera in this part of London except among the persons who were in the habit of drinking the water of the above-mentioned pump well." On the evening of September 7th, Snow gatecrashed a meeting of the parish Board of Guardians, who oversaw public health in the locality, and "represented the above circumstances to them. In consequence of what I said, the handle of the pump was removed on the following day." The epidemic soon petered out. Contrary to popular myth, some versions of which have Snow heroically disabling the pump himself, it was not this action that stopped the epidemic in its tracks; it had already peaked and putting the pump out of action proved nothing. A more detailed analysis would be required to make the case.

The most famous aspect of the affair has become Snow's "death map"—a map of the streets with each cholera fatality marked with a black bar on the building where they lived. He first exhibited such a map in December 1854, and it went on to become a centerpiece of his

NOT A DROP TO DRINK
A satirical cartoon from 1849, commenting on the problems attendant on London's water supply in the course of a cholera epidemic.

THE DEATH MAP

**SNOW'S CHOLERA
MAP OF 1854**
Note his extremely early use
of a bar chart-style approach
to the graphic depiction of
quantitative data, where the
number of deaths at each
household is indicated by
the size of the black bar.

second, expanded edition of *On the Mode of Communication of Cholera*. In fact, Snow was not the first to create such a map, having probably seen an earlier example in T. Shapter's 1849 work, *The History of the Cholera in Exeter* in 1832, and possibly being acquainted with a map of September 1854, very similar to his own eventual effort, produced by Edmund Cooper, an engineer for the Metropolitan Commission of Sewers. Snow called his map a "diagram of the topography of the outbreak"; what made it special was his careful investigation and analysis. The map showed that the vast majority of cases were in close proximity to the Broad Street pump, but Snow went further, showing that even the outlying cases involved people who had drawn their water from the same pump (e.g. an old woman who "preferred" the taste of water from Broad Street). Where there were anomalies, such as the absence of cases at a brewery and workhouse very near the pump, Snow showed that these institutions had their own wells and thus "never sent to Broad Street for water."

In a later version of the map, Snow ingeniously—and "by the most careful calculation"—drew a line marking the median distance between the Broad Street pump and the others that were nearest to it in the district. This proved that the Broad Street pump "catchment" perfectly aligned with the area containing the vast majority of cholera cases.

Despite Snow's strong case, Whitehead's subsequent findings and an excavation of the Broad Street well that showed it did indeed have filth leaking into it from neighboring cesspools, there was official resistance to the waterborne transmission theory. The Board of Health issued a report that said, "we see no reason to adopt this belief" and shrugged off Snow's evidence as mere "suggestions." Snow accepted that he might not see cholera epidemics stopped in his lifetime. Sadly, he was right, for he died just a few years later in 1858 at the age of just 45, but when cholera did next strike London, new sewers were already under construction and official advice was to boil water before use. Snow's work thus did save lives, and after Robert Koch identified *Vibrio cholerae* in 1883 his theory gained widespread acceptance.

THE SOURCE OF THE NILE: BURTON AND SPEKE TRAVEL TO THE HEART OF AFRICA

1858

CATEGORY

Astronomy

Biology

Chemistry

Exploration

Mathematics

Medicine

Physics

Technology

Discoverer: John Speke

Circumstances: Searching for the source of the Nile in Africa, Speke leaves his sick partner at base camp and strikes out on his own, following local rumors of a vast lake

Consequences: Discovers the source of the Nile, opening East Africa to colonization

... the moment he sighted the Nyanza he felt at once no doubt but that the "lake at his feet gave birth to that interesting river which has been subject of so much speculation and the object of so many explorers." The fortunate discoverer's conviction was strong; his reasons were weak...

Richard Burton, *The Lake Regions of Central Africa* (1860)

John Hanning Speke was the first European in at least 1,800 years, and possibly ever, to set eyes upon what he called Victoria Nyanza, the largest freshwater lake in the world and the source of the White Nile. His discovery should have laid to rest a mystery thousands of years old, which had gripped the Victorian imagination, yet it caused controversy and bitterness.

THE MOUNTAINS OF THE MOON

The Nile was sacred to the ancient Egyptians and fascinating to the Greeks and Romans. None of them knew for sure where it rose, although some accounts correctly indicated that it is formed by the confluence of two major streams, later known as the Blue Nile and the White Nile. The influential 2nd-century CE geographer Ptolemy drew upon tales of the remarkable 1st-century Greek trader and explorer Diogenes, who appears to have traveled inland from the East African coast and reached what are now called the Rwenzori Mountains and Victoria and/or other African Great Lakes. These he described as the source of the Nile, and Ptolemy included them on his seminal map, calling the peaks the Mountains of the Moon. A legend was born, and medieval cartographers produced their own copies of Ptolemy's now lost original which included the Mountains of the Moon, although by the medieval period the source of the Blue Nile, in Ethiopia, was known. It was also known that the White Nile extended far to the south, so the mysterious mountains were also shifted south. By the early colonial period it was suspected that the White Nile ran the length of Africa and perhaps started somewhere in South Africa.

In 1770 the Scottish explorer James Bruce reached the source of the Blue Nile. But the far distant source of the White Nile remained a huge mystery, and interest in it grew throughout the 19th century. The Napoleonic adventure in Egypt stimulated exploration, initially by French officers. By 1840 the French had traveled to within four degrees latitude of the Equator, yet the White Nile remained a wide, mature river, indicating that they were still a long way from the source. In an era of imperial adventurers, the discovery of the source of the Nile and the fabled Mountains of the Moon offered one of the great prizes of the age.

Among those with their eyes on the prize was the notorious adventurer, linguist and scholar, Richard Burton. Burton had served in the British

Army in India, but was most famous for daringly disguising himself as a Muslim to make the pilgrimage to Mecca, forbidden to non-Muslims on pain of death. Burton had a taste for adventure, writing in his 1872 book *Zanzibar*, "Of the gladdest moments in human life, methinks, is the departure upon a distant journey into unknown lands." In 1854 Burton launched an expedition to Somali-Land (modern Somalia), probably with an eye to discovering the source of the Nile. When a member of his party died, he drafted in as replacement a young officer on furlough from the Army in India, John Hanning Speke. The distinguished African explorer Samuel Baker later described him as a "painstaking, determined traveler who worked out his object of geographical research without the slightest jealousy of others—a splendid fellow in every way."

The Somalian expedition met with disaster when the British party was ambushed at Berbera, in a battle that may have sown the seeds of ill feeling between the two men. Unfortunately, Speke's pistol jammed and he was taken prisoner, beaten, and stabbed. "I was a miserable looking cripple," he described himself after being rescued, "dreadfully emaciated from loss of blood, and with my arms and legs contracted into indescribable positions, to say nothing of various angry-looking wounds all over my body." Burton too was wounded, a spear thrust through the cheek leaving him with a roguish scar to enhance his disreputability. But did the Berbera incident leave Speke still feeling resentful and defensive, years later?

Evidently Burton felt no bitterness, for in 1856 he invited Speke to join him on a Royal Geographical Society sponsored expedition to East Africa, officially to follow up on reports of "the Sea of Ujiji," a huge lake far inland, but in reality to find the source of the Nile. In December they arrived at Zanzibar; the following June they finally struck inland, and by November 1857 they had reached Kazé. Here they met a fascinating informant, an Arab trader called Sheik Snay, who told them about three inland lakes. Speke started to wonder whether the northernmost one might be the elusive source.

© Popperfoto | Getty Images

SPEKE AND BURTON
At top is a tinted version of a sepia photograph of John Hanning Speke in heroic mode with his explorer's accoutrements—a sextant and shotgun—in front of a studio backdrop showing an imaginary version of Lake Victoria. Below is a photo of Sir Richard Burton; clearly visible is the scar he picked up during the skirmish in Somaliland.

**THIS
MAGNIFICENT
SHEET OF
WATER**

Burton and Speke continued eastward, reaching Lake Tanganyika in February 1858, but the expedition was fraught with difficulty, danger, and disease. Porters deserted them, local chieftains obstructed and extorted them, Speke went temporarily blind, and Burton was so ill he had to be carried everywhere. They struggled back to Kazé, by which time Speke was sufficiently recovered to ask if he could strike out on his own, on what Burton would later call "a flying trip" to investigate the reports of a lake to the north. On July 9th, 1858, Speke set out with 35 men and marched due north for a month. On August 3rd he climbed a hill and "… the vast expanse of the pale-blue waters of the Nyanza burst suddenly upon my gaze… (This magnificent sheet of water I have ventured to name VICTORIA, after our gracious sovereign.)" Nyanza is the Bantu word for "lake," so it became Victoria Nyanza.

Speke took compass bearings and gathered local information, although he had already made up his mind that the lake "gave birth to that interesting river, the source of which has been the subject of so much speculation, and the object of so many explorers." Rejoining Burton on August 25th, he wasted no time in sharing his certainty; Burton was not so sure: "the fortunate discoverer's conviction was strong; his reasons were weak…"

This marked the start of a dispute between the two men. Burton must have been bitter at having missed out, and was still less impressed when he finally got back to England in 1859 to find that Speke, who had hurried back ahead, had already announced his discovery and was claiming all the credit. Burton's main point was that Speke, who was relatively poor at surveying and recording, had insufficient evidence for his claims. Accordingly Speke promptly set off on a second expedition in 1860, reaching the kingdom of Buganda, where he gained permission to travel to the northern outflow of Lake Victoria. On July 28th, 1862, Speke beheld what he called Ripon Falls, the source of the White Nile (or one of them, at any rate).

**THE PROBLEM
OF ALL AGES**

To prove that this was the Nile, Speke needed to follow it until he reached the "known" course. Unfortunately, on reaching Karuma Falls, where the river makes a huge bend to the west, he was obliged by local conditions—there was a war on—to leave the river and cut across. On December 3rd he reached a European ivory trader's outpost, and by May 1863 he was back in Cairo, announcing to the world that the

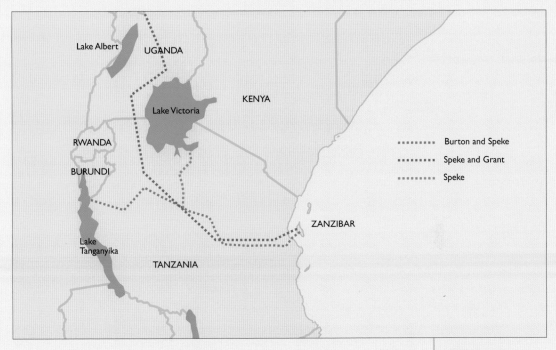

source of the White Nile had been found. He was greeted as a hero on his return to London, but his failure to follow the entire course of the river left him vulnerable to attacks, led by Burton.

A debate between the two men was scheduled for September 16th, 1864, but the day before, Speke was tragically killed in a hunting accident, when his own rifle discharged as he climbed over a wall. There were dark mutterings, possibly instituted by Burton himself, that he had committed suicide rather than face his main detractor in open debate.

In fact, Burton had been wrong and Speke had been right. Lake Victoria is indeed a primary source of the White Nile (alongside springs in the fabled Mountains of the Moon, which Speke had also glimpsed). Speke had solved what Sir Roderick Murchison, president of the Royal Geographical Society, called the "problem of all ages." He had become the first European to cross Equatorial Eastern Africa, filling in a 500-mile (800-kilometer) wide blank space on the map, and opening up both central East Africa and the entire course of the Nile to further exploration, domination, and finally colonization.

VOYAGES TO THE INTERIOR
Map showing Burton and Speke's path to Lake Tanganyika, Speke's solo detour to Victoria, and his follow-up expedition with James Augustus Grant to find the outflow of Lake Victoria.

CATEGORY

Astronomy

Biology

Chemistry

Exploration

Mathematics

Medicine

Physics

Technology

THE LAWS OF INHERITANCE: GREGOR MENDEL'S PRINCIPLES

1863

Discoverer: Gregor Mendel

Circumstances: Seeking to understand the rules governing heredity, an Austrian monk spends years tending pea plants in the shade of the monastery walls

Consequences: Uncovers the algebra of heredity, the basis for the modern genetic synthesis

My scientific studies have afforded me great gratification; and I am convinced that it will not be long before the whole world acknowledges the results of my work.

Gregor Mendel, 1883

Today, the work of Gregor Mendel is celebrated as one of the foundation stones of modern biology. He has even achieved the ultimate accolade: his name has become an adjective. Biology textbooks from high school to postgraduate level speak of Mendelian genetics and Mendelian inheritance. This unassuming Austrian monk ranks alongside Darwin in the pantheon of biology.

A QUESTION OF BREEDING

Mendel, a bright farmer's son who excelled academically but struggled all his life to overcome anxiety and shyness, received his scientific training against the background of raging controversies in the field of heredity. Evolutionary theories were in the air, with debate over how species arose and whether they could transform into new species. How were characteristics—of plants, animals, and humans—passed down from one generation to the next? Did each generation contain preformed miniature versions of itself, like a series of nested Russian dolls, or were the characteristics of parents somehow combined, so that each offspring was a hybrid of its parents? According to recent advances that informed Mendel's thinking, the latter was the case, and he also learned about the new cell theory, which said that the fundamental unit of life was the cell, with the process of reproduction resulting from the combination of germ cells—the male seed (pollen or sperm) and the female seed (the egg).

GREGOR MENDEL
Mendel only became a monk because of financial insecurity, later admitting "my circumstances decided my vocational choice."

Driven by financial insecurity into the haven of monastic life at the Augustinian Convent in Brünn, Mendel was encouraged to pursue researches in plant breeding. He wished to discover whether hybridization involved straightforward blending of the characteristics of parents. Were there rules that governed the process? Did it result from simple "fractional" mixing, so that the color of an offspring plant resulted from a process analogous to mixing ink and water in different proportions? To his study he brought careful, methodical application, mathematical rigor, and a remarkable clarity of thought.

GIVE PEAS A CHANCE

Mendel began by testing different varieties of the edible pea (*Pisum sativum*), learning which characteristics—or traits—were easy to work with, and breeding pure lines of plants: ones that only produced offspring with constant traits. One of the traits he chose was seed color, which could be either green or yellow, and for two years he bred lines

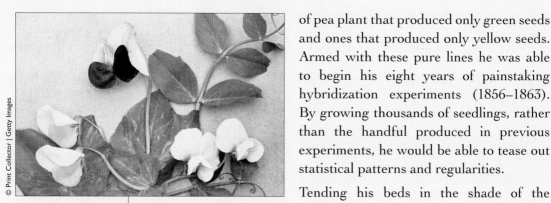

© Print Collector | Getty Images

PRETTY IN PINK
Pea plants showing classic
Mendelian phenotypes,
in the form of white
or pink flowers.

of pea plant that produced only green seeds and ones that produced only yellow seeds. Armed with these pure lines he was able to begin his eight years of painstaking hybridization experiments (1856–1863). By growing thousands of seedlings, rather than the handful produced in previous experiments, he would be able to tease out statistical patterns and regularities.

Tending his beds in the shade of the monastery wall he amassed results, the most striking of which applied to the seed color trait. He bred pure green-seed peas with pure yellow-seed peas to produce hybrid progeny, and from these he harvested 6,022 yellow seeds and 2,001 green seeds, a ratio of almost exactly 3:1. When he then bred the yellow-seed producing hybrid progeny with each other, he discovered that only a third of the second generation bred true (i.e. produced only yellow-seeded third generation plants); the other two-thirds of the second generation plants were obviously hybrids, plants that somehow possessed the ability to produce both yellow and green-seeded progeny, even though they themselves produced only yellow seeds (today this is known as the difference between genotype—genetic characteristics—and phenotype—expressed characteristics). Meanwhile, the green-seeded second generation plants produced only green-seeded progeny. In other words, the true proportions of the second generation plants were 1 quarter pure yellow-seeded, 2 quarters hybrid, and 1 quarter green seeded, or 1:2:1.

Looking at these numbers Mendel, who taught mathematics at the local high school, was struck by their resemblance to what is known as a binomial expansion: the result of the equation $(Aa)^2$, which is $1A^2\ 2Aa\ 1a^2$. A and a were algebraic terms, the language of mathematics, but Mendel now took a revolutionary step and applied them to biology, in combination with what he had learned from cell theory. He used "A" and "a" to represent the potential traits carried in the germ (egg and pollen) cells, with "A" being the dominant potential—the one that obscures the other when both are present; in this case, yellow seeds. Meanwhile, "a" represents the recessive potential: green seeds. He hypothesized that these trait units, which came to be called alleles, are separate in the germ cells, but are brought together through fertilization so that the resulting plant has a pair of them. The three classes of plants he had detected in the ratio 1:2:1 could be represented as 1 *AA*:2 *Aa*:1 *aa*. He further hypothesized that when this progeny produces germ cells of its own, the alleles are once again separated, producing "as many sorts of egg and pollen cells as there are combinations possible of formative elements." His extensive breeding experiments suggested to him that all the traits he studied obeyed this rule independently of one another. These hypotheses, "germinal segregation" and "independent assortment," are now known as Mendel's laws.

In February and March 1865 Mendel presented his findings at two meetings of the Association for Natural Research in Brünn (they were published in the association's proceedings the following year), but was met with polite incomprehension, while few scientists read the proceedings of the obscure society. Why did nobody grasp the implications of Mendel's work? A fatal combination seems to have been to blame: Mendel's own shyness; the revolutionary nature of his work, too far ahead of its time to be understood; the obscurity of the journal in which it was published; and the apparent conflict between his findings and the gathering tide of Darwinism. Mendel, a cleric, saw his results partly as a refutation of Darwinian evolutionary theories—if traits were inherited as invariate units, as his work seemed to show, then new traits, and by extension new species, could not arise (it is now known that mutation can overcome this problem). Yet this was precisely the period when Darwin's theory was sweeping the scientific community, gathering evidence that confirmed its absolute truth. Darwin himself favored fractional blending as the principle of heredity, so Mendel's theories were unfashionable from the start.

On top of all this, the necessary conceptual framework for making sense of Mendel's results was missing. There was insufficient understanding of cell division, germ cell formation, or chromosomes, so there appeared to be no physical basis for his algebraic hypotheses. Not until Mendel's work was rediscovered was this physical context available.

Mendel had bred the seeds of a revolution in biology, but they fell on stony ground. Discouraged by the apathetic response to his years of

INTO THE VOID

COMETH THE HOUR

labor, and sent on a wild goose chase by the preeminent botanist of the day, he switched his focus to monastic life. Yet he was only too well aware of the unheralded significance of his research, commenting in 1883, shortly before his death, "My scientific studies have afforded me great gratification; and I am convinced that it will not be long before the whole world acknowledges the results of my work."

In fact, it was to be 17 years more, but the context of the rediscovery of Mendel's work lends support to the notion that scientific breakthroughs are made when their time has come. In 1900 three different botanists, Hugo de Vries, Carl Correns, and Erik Tschermak, were independently pursuing researches similar to Mendel's efforts 34 years before. None of them was aware of his breakthroughs (although two of them had seen his paper, they had not previously grasped its significance), or of each other, but when they embarked on a survey of preexisting literature prior to publication of their results, each was astonished to find that Mendel had got there first. Backed up by this new confirmation, Mendel's paper was revived, spreading to the English-speaking world via the prestigious Royal Horticultural Society (RHS), which arranged for a translation. Cambridge biologist William Bateson championed Mendel's cause, introducing the term "genetics" to a conference of the RHS in 1906.

Mendel's work led biologists to search for the units of heritability that his discoveries so strongly implied, a search that would lead to the discovery of the DNA code (see page 219). More directly, they made it possible to understand how traits, including many genetic conditions and diseases, are passed on. Modern genetic counseling, which helps couples to manage the risk of passing on genetic disease, is only possible thanks to Mendel.

RADIO WAVES: HEINRICH HERTZ PICKS UP THE FIRST TRANSMISSION

1886

CATEGORY

Astronomy

Biology

Chemistry

Exploration

Mathematics

Medicine

Physics

Technology

Discoverer: Heinrich Hertz

Circumstances: Looking to test Maxwell's prediction that electricity can generate electromagnetic waves, Hertz constructs a transmitter and a receiver

Consequences: Proves that light and electromagnetism are one; discovers radio waves

By means of a series of masterly experiments based upon certain phenomena previously discovered by other scientists, Dr. Hertz, between the years 1886 and 1891, added greatly to the knowledge of these electric waves and their effects on adjacent bodies, enabling them to be put to practical use in wireless telegraphy.

"Hertzian Waves," *Amateur Work* **magazine, November 1901**

Hertz's elegant demonstration of the existence of radio waves is regarded as a landmark in the history of physics, both because of the huge technological applications it engendered—leading to the birth of a swathe of new industries such as telecommunications—and because it offered tantalizing clues to a wholly new type of physics.

MAXWELL LIGHTS THE WAY

In 1864 Scottish scientist James Clerk Maxwell introduced his "Dynamical Theory of the Electromagnetic Field" to the Royal Society in London, explaining that light "… is an electromagnetic disturbance in the form of waves propagated through the electromagnetic field according to electromagnetic laws." Maxwell knew from the work of Faraday on induction that a changing electric field can produce a changing magnetic field. His theory went further, predicting that an accelerating electrical charge will generate oscillating electric and magnetic fields, or electromagnetic waves, and that these will propagate indefinitely (unless absorbed by something) at the speed of light. Maxwell's theory suggests that electromagnetic waves and light are one and the same thing, so that a changing electric field should be able to produce light (in the broader sense of electromagnetic radiation). The challenge for physicists following after Maxwell was to confirm his predictions and unify light, electricity, and magnetism into a single phenomenon.

The first to achieve this in a clear and systematic fashion was Heinrich Hertz in 1886. Born in Hamburg in 1857, Hertz was a brilliant physicist noted for his skill at designing and building his own instruments. As a teenager he built, as a hobby, a spectroscope and a galvanometer that were so well designed he was able to use them for his academic studies for years afterward. Hertz studied at a range of German universities, and was particularly influenced by the great German physicist Hermann von Helmholtz, under whose tutelage he did important research on the nature of electricity.

TRANSMISSION RECEIVED

His greatest prize would come in the years 1886–89, at the Karlsruhe polytechnic, where he designed an elegant apparatus to test Maxwell's predictions about the electromagnetic waves. In order to generate the waves, according to Maxwell's theories, a changing electrical field would be necessary, so Hertz set up a circuit with an induction coil to produce high voltages, a capacitor (a device that stores charge), and a spark gap, a tiny gap between conducting poles (small metal spheres).

He described the circuit like this: "Imagine a cylindrical brass body, 3 cm [1³⁄₁₆in] in diameter and 26cm 10¼in] long, interrupted midway along its length by a spark gap whose poles on either side are formed by spheres of 2cm [¹³⁄₁₆in] radius." When the spark crossed the gap charge would travel back and forth extremely rapidly, or oscillate, and this should generate electromagnetic waves that would radiate outward. This setup constituted a transmitter, sending out electromagnetic waves.

In order to prove that these waves had indeed been transmitted, it was necessary to have a receiver. To do this Hertz exploited two phenomena: electromagnetic induction, where a changing electromagnetic field can induce an electrical current in a wire; and resonance, where waves of the right dimensions can cause a body to vibrate in harmony. The turn of the century hobbyists' magazine *Amateur Work*, in an article on Hertzian waves, gave this common example: "Certain bodies are responsive to a particular rate of vibration. If a violin be played close to a wine-glass in exactly the same tone as the vibration rate of the wine-glass, the wave motion from the violin will set up a vibration in the glass, sometimes so violent as to cause the glass to break in pieces." In Hertz's apparatus, the "body" was a loop of wire that would be induced by the electromagnetic waves to produce an oscillating current that resonated with the transmission waves. The receiver loop also had a spark gap, so that when the transmitted waves induced an oscillating current in the loop it would be visible as sparks between the poles.

On November 11th, 1886, Hertz switched on his apparatus, and saw that when the transmitter loop produced a spark, a smaller spark did indeed jump across the gap in his receiving loop. He was able to produce these sparks over a distance of up to 50 feet (15 meters). Maxwell was right, and Hertz had just discovered that electromagnetic waves could indeed be produced by a changing electrical current. He then used a series of experiments to probe the nature and properties of the waves he had just demonstrated. By varying the position of the receiver and monitoring the production of side sparks around the primary spark, Hertz was able to show that the transmission did indeed have a wave

HEINRICH HERTZ
Noted for his skill in the design of experiments and apparatus, Hertz discovered both radio waves and the photoelectric effect.

© Science & Society Picture Library | Getty Images

**RADIO
INSTRUMENTS**
Various spark gaps and
transmitting devices from
the early history of the
investigation of radio and
"wireless telegraphy."

pattern, and to measure its wavelength. Using a rotating mirror he then determined the frequency of the invisible waves, and using these quantities Hertz could thus calculate its velocity, which turned out to be the speed of light. He went on to show that the waves traveled in straight lines and could be focused, diffracted, refracted, and polarized.

In late 1887 Hertz sent to the Berlin Academy his first report on his discoveries: "On Electromagnetic Effects Produced by Electrical Disturbances in Insulators," followed by updates on his subsequent experiments. He had discovered that Maxwell was right: light and electromagnetism are one and the same phenomenon, and it is possible to use electrical fields to generate electromagnetic waves. The waves he had discovered, with wavelengths in the region $11^{13}/_{16}$–$39^{3}/_{8}$ inches (30–100 centimeters), were originally known as Hertzian waves, but later became more widely known as radio waves. His name does survive, however, as the name of the unit for frequency, the hertz (one cycle per second). The radio waves that Hertz had produced had frequencies in the range 1,000–300 megahertz, but the full spectrum of radio waves ranges from 30 kilohertz to 3 gigahertz, with wavelengths from the length of a football to greater than the diameter of the Earth. The discovery of radio waves soon led to applications from telecommunications to astronomy. As early as 1890 Thomas Edison proposed that radio waves might be detected from the Sun.

THE PHOTOELECTRIC EFFECT

The spark in the receiver was so small that Hertz had difficulty seeing it, but his attempt to overcome this problem led him to notice "A phenomenon so remarkable [that it] called for closer investigation." Further experimentation revealed that it was ultraviolet light being given off by the transmitter spark that was responsible. Hertz had discovered the photoelectric effect, but did not venture to speculate on its causes. Later scientists would show that the photoelectric effect is caused when light excites electrons and causes them to fly off, but that this only happens in a way that suggests the light is composed of small packets or particles of energy, which are today called photons. So, Hertz's research on light waves suggested that light is both wave-like and particle-like in nature, a discovery that helped lead to quantum physics.

THE POLYPHASE MOTOR: TESLA SWITCHES ON THE POWER AGE

1881

Discoverer: Nikola Tesla

Circumstances: Tesla observes the inefficient use of commutator brushes to convert alternating current to direct current, and investigates a way to do without the brushes

Consequences: Invents the polyphase motor, making the AC electricity system possible

… the idea came like a flash of lightning and in an instant the truth was revealed. I drew with a stick on the sand, the diagram shown six years later in my address before the American Institute of Electrical Engineers… I cannot begin to describe my emotions.

Nikola Tesla, *My Inventions: The Autobiography of Nikola Tesla* (1919)

The name of Tesla should be as well known as those of Edison and Faraday, if not Newton and Einstein. He was the greatest genius of electrical science and the creator of the basic architecture of modern electricity infrastructure; the father of the Power Age. His biographer Margaret Cheney describes him as "the most versatile and productive genius in history."

THE BRUSHES GAVE TROUBLE

Nikola Tesla was born in 1856, of Serbian parents in the time of the Austro-Hungarian Empire. He had a strange and difficult childhood, surviving dangerous illnesses, having some odd experiences, and dreaming of fantastic inventions and incredible machines, such as a system for blasting message containers around the world through water filled tubes, or a giant space hoop circling the Earth at the Equator. He even dreamed of a colossal waterwheel that would harness the power of Niagara Falls.

NIKOLA TESLA
Remarkable for both his prolific genius and eccentricities, Tesla was of Serbian descent but spent most of his career in America.

Attending the polytechnic at Graz, he saw demonstrated a Gramme dynamo, an electrical generator that could function in reverse as a motor. In order to function with direct current (DC), the type of electrical current produced by a battery, and the type that would form the basis for the earliest attempts to create an electricity system, the Gramme dynamo needed a device called a commutator. The commutator connected to the rest of the device via brushes, but when the dynamo was run in reverse as a motor, "the brushes gave trouble, sparking badly"—and wasting energy.

Tesla had an epiphany. If the dynamo were run with alternating current (AC) the commutator could be dispensed with and the machine's efficiency would be much greater. When he proposed this, the professor explained at length why this could never work—nobody had succeeded in getting AC to drive a motor, and nobody could.

Why did Tesla's teacher believe this, and why was AC deemed inferior to DC? A bit of background in electricity is necessary to understand why. Moving a conductor, such as a copper wire, through a magnetic field generates an electrical current in the wire, through the process known as induction. Induction had been discovered by Michael Faraday in 1831, opening the door to a new world of electricity generation and application. In a generator, a source of power, such as falling water or steam produced by heating water by burning oil or coal, spins a coil of wire through a magnetic field and induces a stream of electricity. The principle also works in reverse, so that a current running through a wire generates a magnetic field. Magnetic fields have a polarity, a north and south, and while opposite polarities attract, similar ones repulse one another. This means that by using a current to create a magnetic field, it is possible to attract another magnet (such as one attached to the rotating arm of a motor), converting electrical energy into motion and driving a motor.

Whereas a battery produces electrical current that flows in one direction only (direct current, or DC), a generator like the one described above, where a coil of wire spins through a magnetic field, produces current that switches direction back and forth very rapidly, and this is alternating current, or AC. At the time of Tesla's encounter with the Gramme dynamo, AC current was largely considered to be an irritating curio. Initial attempts to drive a motor with AC had foundered because the fields generated reversed several times a second, and therefore so did the direction of motion they could induce—the net result was that motions back and forth canceled each other out. The solution adopted was to convert AC into DC by means of mechanical switches, in the form of the inefficient and sparky commutator that caught Tesla's attention in the Gramme dynamo. DC was simpler to understand and therefore seemed simpler to work with, even though it had drawbacks.

Tesla knew that there was a way to make AC work, if he could only figure it out. The problem occupied him for several years until in 1881, while walking with a friend in the park at Graz and reciting lines from Goethe's *Faust*, he was struck by a blinding flash of insight:

LIKE A FLASH OF LIGHTNING

> As I uttered the inspiring words the idea came like a flash of lightning and in an instant the truth was revealed. I drew with a stick on the sand, the diagram shown six years later in my address before the American Institute of Electrical Engineers, and my companion understood them perfectly. The images I saw were wonderfully sharp and clear and had the solidity of metal and stone, so much so that I told him, "See my motor here; watch me reverse it." I cannot begin to describe my emotions... A thousand secrets of nature which I might have stumbled upon accidentally, I would have given for that one which I had wrested from her against all odds...

The device he had sketched in the sand was the polyphase AC motor, Tesla's first major invention, and the one he is best remembered for by the scientific mainstream. His inspiration was to produce a motor that did not require a physical connection like the brushes of the commutator. Instead, he used two currents that were out of step to induce magnetic fields in coils of wire arranged around the outside of a circular magnetic motor element attached to a rotating shaft. As one phase of current was dying away the other was peaking, so that there was always a strong current to power the motor. Because the peaks followed each other successively around the coils in their circular arrangement, it was possible to make the attractive magnetic field move constantly round in a circle. The inner magnetic element would always be chasing this field, producing constant circular motion. This motor, which used more than one phase of current and hence was known as a polyphase motor, was much more efficient and reliable than the DC motor.

Crucially, it also meant that it would now be feasible to supply electricity to people's homes in AC form, and they would be able to use it directly to power their domestic devices. This had the potential to transform the market for electricity: AC can be transmitted much farther than DC because it can be transformed to high voltages (good for transmission) and then back to low voltages (good for domestic use). Tesla would go on to invent generators for making AC and transformers for changing its voltage. It was his breakthrough that allowed electricity to be generated, distributed, and used on a massive scale, ushering in what has been called the Power Age.

SEVEN-LEAGUE BOOTS

Over the next two months Tesla conceived a series of inventions for the production, distribution, and use of AC power, including generators, transformers, and motors—everything that was needed to set up an entire infrastructure of AC electricity, which would be far superior to the DC system being promoted in the US by Thomas Edison (Tesla's system still constitutes the basis of the modern electricity network). Tesla recalled giving himself up "entirely to the intense enjoyment of picturing machines and devising new forms… When natural inclination develops into a passionate desire, one advances toward his goal in seven-league boots."

In developing these many new inventions, Tesla put to good use his remarkable powers of visualization. In *My Inventions* he wrote, "When I get an idea, I start at once building it up in my imagination. I change the construction, make improvements, and operate the device in my mind. It is absolutely immaterial to me whether I run my turbine in my thought or test it in my [work]shop... Invariably my device works as I conceived that it should, and the experiment comes out exactly as I planned." Although in later life he was forced to rely more on real-world trial-and-error experimentation, this virtual reality design system stood him in good stead, for he had an astonishingly good record of creating successful prototypes of his inventions.

© Science & Society Picture Library | Getty Images

POLYPHASTIC!
Tesla's polyphase induction motor: alternating currents in the outer windings (known as the stator because it does not move) generate rotating magnetic fields that cause the central rotor element to spin.

Despite the innate superiority of Tesla's AC power system, with the polyphase motor at its heart, there would be an almighty battle before it won out over DC. Tesla and his backer, the inventor and industrialist George Westinghouse, found themselves facing an implacable and ruthless opponent in the shape of Edison in what became known as the War of the Currents. Edison used dirty tricks to spread fear and disinformation about AC current, claiming via a stooge that it was a "damnable death current," and sponsoring a series of ghoulish animal electrocutions by AC, culminating in the horrific killing of an elephant and the botched execution of the first victim of the electric chair. The Tesla-Westinghouse system was better and cheaper, however, and by 1893 Edison had lost the War of the Currents. That same year Tesla's childhood dream came true when Westinghouse won the contract to build the new hydroelectric plant at the foot of Niagara Falls. Tesla designed and oversaw construction of three massive turbines.

CATEGORY

Astronomy

Biology

Chemistry

Exploration

Mathematics

Medicine

Physics

Technology

X-RAYS: ROENTGEN DISCOVERS A MYSTERIOUS RADIATION

1895

Discoverer: William Roentgen

Circumstances: Experimenting with a cathode ray tube, Roentgen discovers that a fluorescent screen on the other side of the room is glowing

Consequences: Discovers X-rays, leading to medical radiography and radiotherapy

For nowadays, I hear they'll gaze,
Thro' cloak and gown—and even stays;
These naughty, naughty Roentgen Rays.

Popular verse (1896)

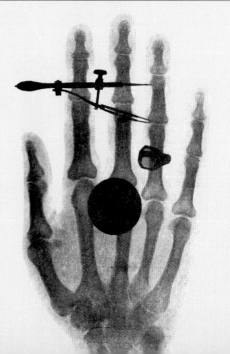

The end of the 19th century and the early years of the 20th century saw a sudden explosion of discoveries in physics, as a whole new world of atomic and subatomic physics opened up. One of the first and most celebrated, thanks to its remarkably rapid real-world impact, was the characterization of X-rays. According to the American Physical Society: "Few scientific breakthroughs have had as immediate an impact as Wilhelm Conrad Roentgen's discovery of X-rays."

"If only there was some way of making the human body transparent like a jellyfish!" exclaims a young doctor faced with a difficult diagnosis, in an 1892 story by the German writer Philander (aka Ludwig Hopf). The protagonist's wish is granted with the appearance of Electra, a young woman surrounded by light, who, "for the good of men and women," gives him a box that emits magic rays, capable of revealing the inner workings of the body. The doctor manages to recreate artificially the properties of the magic light, disseminating his life-saving discovery and declaring, "A new and glorious age has dawned for us physicians." This remarkably prescient story would shortly come true in almost every particular, with the notable exception of Electra herself.

The magic light of Electra had in fact already been accidentally discovered several times, yet no one had noticed or investigated properly. What came to be known as X-rays are high frequency electromagnetic radiation, of shorter wavelength and higher energy than ultraviolet light, with wavelengths ranging from 0.03 to 3 nanometers, around the diameter of an "average" size atom. They can be produced by a phenomenon known as the Bremsstrahlung ("braking radiation") effect, which is where a high-speed electron hurtling toward an atom is rapidly decelerated by the negative charge of the atom's electrons, which repel it. As it slows down it loses kinetic energy, and, following the law of conservation of energy, this energy cannot simply disappear. Instead it is given off as a high-energy photon, an X-ray, which is ejected at right angles to the course of the decelerating electron.

To produce the Bremsstrahlung effect, electrons must first be separated from atoms and then accelerated to high speeds, and this had been happening since 1838. In that year Michael Faraday fitted electrodes inside either end of a glass tube and pumped out most of the air. Passing an electric current through the "rarefied" air resulted in a

THE AGE OF ELECTRA

slight glow. In the 1850s it became possible to achieve "harder" (i.e. much lower pressure) vacuum, so that virtually every molecule of air had been pumped out, and it was found that although the glow in the tube disappeared (because there were no air molecules to excite), the glass at the end of the tube opposite the cathode began to glow instead. Evidently rays of some sort must be coming from the cathode and striking the glass at the far end, and these were known as cathode rays. The English scientist William Crookes achieved the hardest vacuum yet and his "Crookes tube" became the standard apparatus for investigating cathode rays. Crookes himself famously demonstrated that the rays could cast a "shadow" by constructing a tube with a Maltese cross-shaped plaque at one end, and also made a tube with a paddle wheel apparatus, which turned as the cathode rays fell on it, suggesting that the ray consisted of a stream of particles.

Many of these vacuum tube experiments must inadvertently have generated X-rays through the Bremsstrahlung effect, and indeed Crookes himself had noticed that his tubes could produce fogged spots on photographic plates in his laboratory. Several others also observed similar effects, yet no one investigated systematically until 1895.

REMARKABLE RAYS

The man who finally gave X-rays to the world was the German physicist Wilhem Conrad Röntgen (commonly spelled Roentgen in English). Born in 1845, Roentgen became a distinguished scientist despite having severe setbacks, such as being kicked out of technical school in 1862 after being blamed for drawing a caricature of one of the teachers. In fact it was another pupil who was to blame, but the failure to gain his diploma would hamper Roentgen's early career and force him to attend university in Zurich. Nonetheless, by 1888 he held the chair of physics at the University of Würzburg, in Bavaria, and it was here that he made his great discovery.

On the evening of November 8th, 1895, Roentgen was experimenting with a Crookes tube in a darkened room. "If one passes the discharges of [an] induction coil through a… Crookes tube," he wrote in his report on the studies, "and if one covers the tube with a rather closely fitting envelope of thin black cardboard, one observes in the completely darkened room that a piece of paper painted with barium platinocyanide lying near the apparatus glows brightly or becomes fluorescent." The barium platinocyanide paper acted as a fluorescent screen, akin to the screen of a cathode ray tube television; does its presence indicate that Roentgen knew what he was looking for, or was it there by accident? The report does not state.

The fluorescing screen was on the other side of the room, indicating to Roentgen that the cathode rays themselves could not be responsible for the transmission of energy, even if they were somehow getting out of the tube and through the cardboard. So what was responsible? "For the sake of brevity," he wrote, "I should like to use the term 'rays,' and to distinguish them from others I shall use the name 'X-rays.'" Here Roentgen was borrowing from algebra, where the convention is to denote the unknown with the letter "x."

When asked what his thoughts were at the moment of discovering X-rays, Roentgen replied, "I didn't think, I investigated." He spent the next seven weeks investigating, and discovered that the X-rays can pass, more or less unhindered, through a remarkable range of materials. Crucially, however, some substances, especially more dense ones, are more opaque to X-rays. Roentgen used photographic plates to take X-ray photographs of various materials, most famously, his wife's hand complete with wedding ring. "If one holds the hand between the discharge apparatus and the screen, one sees the darker shadows of the bones within the much fainter shadow picture of the hand itself." The medical applications must have been apparent.

WILLIAM ROENTGEN
Despite being expelled from school for someone else's prank, Roentgen went on to become a distinguished scientist, achieving two of the ultimate accolades: a Nobel prize and becoming a noun.

On December 28th he submitted his "Preliminary Communication," titled "On a New Kind of Rays," and in January 1896 he made his first public presentation, to the Würzburg Physico-Medical Society. An anatomist who assisted him by having an X-ray photograph taken of his hand proposed that the new discovery be named "Roentgen's Rays," and although the name X-ray won out, terms like roentgenization and roentgenograms did prosper for a while. Roentgen would go on to win the first ever Nobel Prize for physics in 1901, "in recognition of the extraordinary services he has rendered by the discovery of the remarkable rays."

With remarkable speed the X-ray went from discovery to application. A front page article in the *New York Sun* on January 26th, 1896, proclaimed:

SUCH PROMPT RECOGNITION

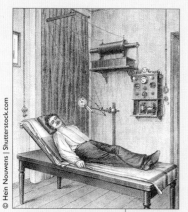

**OLD FASHIONED
X-RAY**
Early radiography setup,
with the patient lying on a
photographic plate while
a sort of X-ray light bulb
suspended above him
blasts out X-rays.

Never in the history of science has a great discovery received such prompt recognition and has been so quickly utilized in a practical way as the new photography which Professor Roentgen gave to the world only three weeks ago. Already it has been used successfully by European surgeons in locating bullets and other foreign substances in human hands, arms, and legs and in diagnosing diseases of the bones in various parts of the body.

In Dartmouth, New Hampshire, for instance, local photographer Howard Langill read about the discovery and promptly contacted his local physics department to suggest that if they had tubes that would generate X-rays, he would supply the photographic plates. After X-ray photographing objects such as a box of weights, one of the physicists contacted his brother, a physician, and on February 3rd a teenager named Eddie McCarthy, who had broken his arm ice-skating, became the first patient diagnosed by X-ray in the US.

Within a year of Roentgen's announcement, the application of X-rays to diagnosis and therapy was an established part of the medical profession. One of the first radiology departments in the world was set up at the Glasgow Royal Infirmary in 1896, where the head of the department, Dr. John Macintyre, achieved a number of firsts: the first X-ray of a kidney stone; an X-ray showing a penny in the throat of a child; and an image of a frog's legs in motion. X-rays also proved potent in attacking tumors and even in treating skin lesions, although it would be a while before it was appreciated that this was because the rays were dangerously energetic. X-rays are a form of "ionizing radiation," meaning they can damage cells and cause genetic mutations, leading to cancer.

The new rays had an immediate impact on the public consciousness. There was widespread alarm at the prospect of a ray that made it possible to see through a person's clothing. Gag companies marketed "X-ray specs," which were completely bogus, but public concern was such that in New Jersey, for instance, a bill was introduced "prohibiting the use of X-rays in opera glasses in theaters." Private detectives speculated on the application of X-rays for spying on cheating husbands, while garment manufacturers produced lead underwear to thwart the X-ray voyeur.

© Hein Nouwens | Shutterstock.com

RADIOACTIVITY: THE HIDDEN POWERS OF URANIUM REVEALED

1896

CATEGORY

Astronomy

Biology

Chemistry

Exploration

Mathematics

Medicine

Physics

Technology

Discoverer: Henri Becquerel

Circumstances: Becquerel investigates whether X-rays are given off when uranium salts fluoresce in sunlight

Consequences: Discovers radioactivity

The discovery of the phenomena of radioactivity adds a new group to the great number of invisible radiations now known, and once more we are forced to recognize how limited is our direct perception of the world which surrounds us, and how numerous and varied may be the phenomena which we pass without a suspicion of their existence until the day when a fortunate hazard reveals them.

Marie Curie, "Radium and Radioactivity," *Century Magazine* **(January 1904)**

Roentgen's discovery of X-rays caused a worldwide sensation and directly led to the discovery of radioactivity. Although in hindsight this can be seen to be a finding of even greater significance than the existence of X-rays, its epochal impact was not evident until some years later, thanks to the research of Marie and Pierre Curie and Ernest Rutherford.

BEYOND THE EXPECTED

Among the scientists astonished and electrified by Roentgen's new rays was French mathematician Jules-Henri Poincaré, who promptly replicated some of the German's X-ray photographs and presented them on the evening of January 20th, 1896, at the weekly meeting of the French Academy of Science in Paris. Pointing out that the X-rays appeared to emanate from the end of the Crookes tube, when it was hit by the cathode ray and fluoresced, Poincaré speculated that perhaps other luminescent bodies might also emit X-rays. (Luminescence includes both fluorescence, the emission of light under stimulation by an external energy source, and phosphorescence, where the light continues to be emitted after the external stimulus ceases.) In the audience was one of the world's leading authorities on luminescent substances, Henri Becquerel.

HENRI BECQUEREL
Like his father before him, Henri was Professor of Physics at the Museum of Natural History in Paris, and an expert on luminescence, the emission of light by materials after exposure to radiation.

Becquerel belonged to a line of distinguished scientists, occupying the same chair of physics at the Paris Museum of Natural History previously held by his father Alexandre Edmond, an expert on phosphorescent minerals. Along with his intellectual heritage, Becquerel senior had left a vital legacy to his son: a collection of uranium salts, noted for their property of very brief phosphorescence after illumination.

Becquerel set to work the very next day. He was well practiced at working with photographic plates, and worked through a series of trials in which he covered a plate with thick black paper, and then fixed on top of the paper a phosphorescent substance, closing it all up in a case. The case would then be exposed to the Sun, the rays from which could not penetrate directly to the plate but which would excite the phosphorescence, which in turn might give off X-rays, Bequerel hoped. He had no success with most of the samples he tested, but, as he recalled in his Nobel Prize lecture of 1903, "… notwithstanding the negative experiments with other bodies, I placed great hopes in experimentation with uranium salts, whose phosphorescence I had formerly studied, following the works of my father. Those bodies… seem to have a particularly remarkable molecular constitution, at least from the point of view of absorption and phosphorescence."

The results were encouraging, as he related to the Academy on February 24th. Uranium salts produced distinct fogging of the plates. "One must conclude from these experiments that the phosphorescent substance in question emits rays which pass through the opaque paper and reduces silver salts." On March 2nd, just a day after his next experiment, Becquerel was back in front of the Academy to report on something "quite important and beyond the phenomena which one could expect to observe." Theorizing that more intense phosphorescence might produce greater emissions, he had prepared further salt-plate cases, hoping to expose them to bright sunlight. However, he explained, "since the sun was out only intermittently on these days, I kept the apparatuses prepared and returned the cases to the darkness of a bureau drawer, leaving in place the crusts of the uranium salt." When the sun still did not shine, he decided to develop the plates anyway, "expecting to find the images very weak. Instead the silhouettes appeared with great intensity. I immediately thought that the action had to continue in darkness…"

Becquerel repeated his accidental experiment under more controlled conditions, assembling cases with uranium salts in direct contact with the plates, and separated by sheets of glass and aluminum, and taking care to shield the cases from even the slightest exposure to sunlight. After a few hours in the darkness, all three of the plates were blackened. "I am now convinced that uranium salts produce invisible radiation, even when they have been kept in the dark," he wrote in his lab journal.

This radiation was initially known as Becquerel rays, taking its place amid a flood of new rays and radiations competing for attention, including radio waves, cathode rays, and X-rays. Becquerel's rays were among the least impressive of the bunch, producing much less distinct "shadow photographs" than Roentgen's and thus having no obvious medical application. Although he published seven papers on the new phenomenon in 1896, Becquerel managed only two papers in

1897 and by 1898 he had lost interest. As Marie Curie later noted, "The world, attracted by the sensational discovery of Roentgen rays, was less inclined to astonishment."

TRANSMUTATION OF THE ELEMENTS IS POSSIBLE

That same year, however, would see the full significance of Becquerel's discovery begin to dawn, thanks largely to the work of his former doctoral student, Marie Curie. Together with her husband Pierre, Curie took a strong interest in uranium and thorium, another element found to exhibit the same property, which Curie termed "radioactivity," on the basis that they actively emit radiation. Becquerel had used photographic plates to detect radiation, but this was a relatively crude tool that did not permit proper quantification of the phenomena. Curie made use of the fact that radiation discharges an electroscope,

BECQUEREL IN HIS LABORATORY
Here we see him experimenting with a large magnet, during his follow-up research into the phenomenon of radioactivity that he had discovered.

enabling her to make precise measurements of intensity of radiation and compare substances as to their radioactivity. The radioactivity of uranium, she discovered, is independent of temperature or other environmental factors. Writing in *Century Magazine* in 1904, she recalled, "Thus I reached the conviction that the emission of rays by the compounds of uranium is a property of the metal itself—that it is an atomic property of the element uranium independent of its chemical or physical state."

She was particularly struck by the finding that "a specimen of pitchblende (uranium ore) was… four times more active than oxide of uranium itself. This observation astonished me greatly. What explanation could there be for it?… The answer came to me immediately: The ore must contain a substance more radioactive than uranium and thorium, and this substance must necessarily be a chemical element as yet unknown." In fact the Curies discovered two new radioactive elements, radium and polonium.

After heroic labors they managed to obtain a tiny amount of pure radium, which proved to be intensely radioactive, generating so much heat it is able to melt its own weight in ice in an hour. "When we reflect that radium acts in this manner continuously," she pointed out, "we are amazed at the amount of heat produced, for it can be explained by no known chemical reaction…" The inevitable conclusion was that, "radioactivity is a property of the atom of radium; if, then, it is due to a transformation this transformation must take place in the atom itself. Consequently, from this point of view, the atom of radium would be in a process of evolution, and we should be forced to abandon the theory of the invariability of atoms, which is at the foundation of modern chemistry... proof will exist that the transmutation of the elements is possible."

Also in 1898 the New Zealand born British physicist Ernest Rutherford began investigating radioactivity. Like Curie he made quantitative measurements, looking at the penetrating power of the radiation emitted by uranium. Interposing aluminum sheets of varying thickness between the uranium and the measuring device, he found that the amount of radiation that is absorbed by the aluminum does not change uniformly as the thickness is increased, but drops suddenly once the thickness is more than a few hundredths of a millimeter. This clearly indicates that the uranium radiation consists of two types of ray, one that is easily blocked, which he called alpha radiation, and one that is more penetrating, which he called beta radiation.

ALPHA, BETA, AND GAMMA

This flurry of research prompted Becquerel to return to the field he had created, and he made several important discoveries, showing that beta radiation can be deflected by electric or magnetic fields, unlike X-rays. In this way Becquerel was able to determine the charge to mass ratio of the beta particle, showing it to be identical to the electron, recently identified in just this fashion by J. J. Thomson (see page 190). In 1903 Becquerel shared the Nobel Prize for physics with the Curies for their discoveries in the field of spontaneous radioactivity.

CATEGORY

Astronomy

Biology

Chemistry

Exploration

Mathematics

Medicine

Physics

Technology

THE ELECTRON: DISSECTING THE CATHODE RAY

1897

Discoverer: Joseph John Thomson

Circumstances: Thomson devises an elegant experiment to measure the charge:mass ratio of what he believes to be tiny particles

Consequences: Discovery of the electron; first proof of subatomic particles

Could anything at first sight seem more impractical than a body which is so small that its mass is an insignificant fraction of the mass of an atom of hydrogen?

J. J. Thomson

Thomson's work on the electron is an example of how science had become more collaborative and incremental in nature, with "discoveries" resulting from the accumulation of research by several people, culminating in a careful demonstration deemed to be decisive.

Born in 1856, English scientist Joseph John Thomson only became a physicist by accident, as his family could not afford to apprentice him to an engineer, as originally intended. Instead Thomson went to a school that excelled in science, becoming a brilliant mathematical physicist and Cavendish Professor of Experimental Physics at Cambridge at the age of just 27, despite having limited practical experience. Although he was clumsy with the instruments he was adept at designing elegant and rigorous experiments. In the 1890s he, like many contemporaries, was interested in the mysteries of the cathode ray. "There is no other branch of physics," he wrote in 1893, "which affords us so promising an opportunity of penetrating the secret of electricity."

At this time there was a fierce debate over the nature of cathode rays: are they streams of particles (aka corpuscles) or waves in the ether (the invisible medium of the universe, through which electromagnetic waves such as light were assumed to travel)? "The most diverse opinions are held as to these rays…" Thomson reflected, "It would seem at first sight that it ought not to be difficult to discriminate between views so different, yet experience shows that this is not the case…"

In Germany Herz had found two strong pieces of evidence supporting the view that the cathode ray is a wave. First he had shown that cathode rays could pass through gold foil; gold is dense, so it seemed implausible that a stream of particles could pass straight through it. Second, he had tested whether cathode rays, which were known to be deflected by magnetic fields, can also be deflected by electrical fields. If they were charged particles, as the particle theory suggested, they should be deflected, but Herz found no evidence of deflection.

Thomson found these results puzzling, probably because he was already largely convinced that the particle theory was the right one. Reflecting on Herz's gold foil findings, he recalled in his Nobel Prize

PENETRATING THE SECRET OF ELECTRICITY

J. J. THOMSON
Despite being notoriously maladroit with experimental apparatus, to the point where assistants begged him not to handle the equipment, Thomson became a brilliant experimental physicist.

lecture of 1906: "The idea of particles as large as the molecules of a gas passing through a solid plate was a somewhat startling one, and this led me to investigate more closely the nature of the particles which form the cathode rays."

He carried out experiments of his own. First he showed that using a magnetic field to bend the cathode ray into a metal container results in the container becoming negatively charged, proving that the ray is definitely carrying the negative charge. Next he repeated Herz's electrical field experiment, which he suspected had not worked because there was an insufficient vacuum in the Crookes tubes the German was using. Any gas left in the tube becomes electrically charged due to the passage of the cathode ray, and the charged gas acts like an insulator around a wire, protecting the cathode ray from the influence of external electrical fields. Thomson showed that with a hard enough vacuum in the tube, the ray can after all be deflected by an electric field. These two findings led Thomson to disagree with the German wave theorists: "I can see no escape from the conclusion that [cathode rays] are charges of negative electricity carried by particles of matter. What are these particles? Are they atoms, or molecules, or matter in a still finer state of subdivision?"

Other researchers had already found clues. Herz's student Phillip Lenard, extending the gold foil experiments to measure how far a cathode ray can penetrate various gases after passing through the foil, had showed that if the ray were composed of particles, they must be incredibly tiny. Meanwhile the German researcher Emil Wiechert, in January 1897, had obtained a measurement of the ratio between charge and mass of the supposed cathode ray particles, finding it to be over a thousand times greater than the ratio for the smallest charged atom.

MEASURING THE UNMEASURABLE

Thomson now repeated Wiechert's measurement with precision and clarity. The task facing him was to seek a hypothetical particle that many scientists at the time did not believe existed, and for which there was no direct means of observation or measurement. In particular, he wanted to work out the probable mass of the particle from the ratio between its charge and mass. There was no way to measure the charge of the particle directly, but if he could measure its velocity he could find the ratio. Unfortunately this could not be directly measured either.

Instead Thomson cleverly used the fact that, for an object in both a magnetic and an electric field, when the forces on a charged object are balanced, the ratio between them is equal to the velocity of the object.

He devised a Crookes tube where both an electrical and a magnetic field acted on the cathode ray; the position of the green glow produced when the ray struck the glass at the end of the tube showed the degree of deflection of the ray. He altered the strength of the fields until the ray was not deflected one way or another, so that the glow was directly at the end of the tube, indicating that the magnetic and electric fields were in equilibrium. Since he knew the strengths of the fields at this point, he could work out the velocity of the charged object, which proved to be "⅓ the velocity of light, or about 60,000 miles [96,500 km] per second." Although this is "many thousand times the average velocity with which the molecules of hydrogen are moving," suggesting a particle smaller than hydrogen, the smallest atom, it is, significantly, much slower than light, suggesting that the cathode rays are not some form of electromagnetic wave.

ELECTRON DETECTIVE
The cathode ray tube apparatus Thomson designed to measure the charge:mass ratio of the electron. The cathode at the right-hand end of the tube generates a beam of electrons and the electromagnet coils and charged plates in the middle generate magnetic and electric fields to deflect the beam.

The next step was to use this calculation of velocity to work out the charge:mass ratio. Turning off the magnetic field caused the cathode ray to deflect; measuring the degree of deflection allowed Thomson to plug in his results to an equation similar to that used to work out the path of a bullet fired from a gun as it falls under gravity. Comparing the ratio thus obtained with the ratio for the smallest particle then known, a hydrogen atom, Thomson found that "for the corpuscle in the cathode rays the value of charge:mass is 1,700 times the value for the corresponding quantity for the hydrogen atom." Furthermore, this ratio proved to be independent of many factors: the velocity of the ray, the source of the ray (Thomson used cathodes of various different metals to generate the cathode rays) or the kind of gas in the tube, strongly suggesting it to be some sort of elementary particle.

What did this mean? Either the "corpuscle" had a charge 1,700 times greater than a hydrogen atom, which was implausible, and which Thomson would soon demonstrate not to be the case, or it had a mass 1,700 times smaller. "We are driven to the conclusion," he told his Nobel lecture audience, "that the mass of the corpuscle is only about 1/1,700 of that of the hydrogen atom. Thus the atom is not the ultimate limit to the subdivision of matter; we may go further and get to the corpuscle, and at this stage the corpuscle is the same from whatever source it may be derived."

PULLING THEIR LEGS

Reaction to Thomson's 1897 findings was skeptical. "At first," he later recalled, "there were very few who believed in the existence of these bodies smaller than atoms. I was even told long afterward by a distinguished physicist who had been present at my lecture at the Royal Institution that he thought I had been 'pulling their legs.'" But Thomson believed that he had found the basic component of matter, "one of the bricks of which atoms are built up." The word "electron" had been in use to refer to charge-carrying entities since 1891, and it now became attached to Thomson's "corpuscle."

Thomson quickly saw the immense explanatory power of the electron, correctly suggesting that the arrangement of electrons in an atom could explain its chemical properties and underlie the structure of the Periodic Table of elements, and that ions become charged by losing and acquiring electrons. However, Thomson was wrong about some things. He believed the electron to be the only subatomic particle, proposing the "plum pudding" model of the atom in which electrons are embedded in some sort of cloud, field, or "sticky" mass of positive charge. This concept of the atom would shortly be shattered by the amazing findings of Thomson's one-time student, Ernest Rutherford (see the next entry).

MAKING SENSE OF MATTER: THE STRUCTURE OF THE ATOM

1909

Discoverer: Ernest Rutherford, Hans Geiger, and Ernest Marsden

Circumstances: Finding unexpected scattering in the passage of alpha particles, Rutherford suggests that Geiger and Marsden look for really big deflections

Consequences: Discovery of the nucleus; founding of nuclear physics

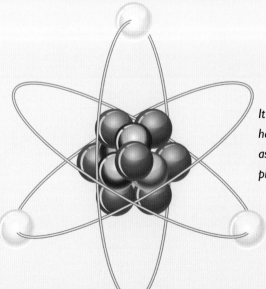

It was quite the most incredible event that has ever happened to me in my life. It was almost as incredible as if you fired a 15-inch [38-centimeter] shell at a piece of tissue paper and it came back and hit you.

Ernest Rutherford on learning the results of the Geiger-Marsden experiment

Ernest Rutherford is the only scientist to have achieved his greatest discovery after winning a Nobel Prize. Thanks to the research he directed, his team members Hans Geiger and Ernest Marsden discovered that alpha particles did not pass through gold foil, as expected, but deflected off at apparently impossible angles, as if they had bounced off something small but very dense.

A RABBIT FROM THE ANTIPODES

Widely regarded as the greatest physicist of his era, Rutherford came from the edge of the world to dominate the capitals of scientific research. Born in 1871 on a farm in New Zealand, Rutherford overcame near poverty and limited opportunities to win a succession of scholarships that propelled him from a rural backwater to becoming, in 1895, the first research student in the history of Cambridge University (i.e. the first person to do a doctorate who had not already studied there). Working under J. J. Thomson at the Cavendish Lab, Rutherford soon impressed. "We've got a rabbit here from the Antipodes and he's burrowing mighty deep," remarked Andrew Balfour, a contemporary; Rutherford would eventually burrow deeper than almost anyone in 20th-century science.

He was ambitious, writing to Mary Newton, his sweetheart in New Zealand: "I have some very big ideas which I hope to try and these, if successful, would be the making of me. Don't be surprised if you see a cable some morning that yours truly has discovered half-a-dozen new elements, for such is the direction my work is taking." Indeed, after initial experiments with radio, where he managed to extend the range of radio broadcasting to up to a mile, Rutherford did turn his attention to new elements: radioactive ones. In 1898 he showed that there were at least two types of radiation, alpha and beta (see page 189), and in 1901 he demonstrated the existence of a third, an electromagnetic wave he called the gamma ray.

Taking up a chair in physics at Canada's McGill University, Rutherford worked with chemists to show that radioactive elements transmuted into new elements as they gave off radioactive particles. This work would help to win him the 1908 Nobel Prize for chemistry. Investigating the properties and behavior of alpha particles, he worked out that they were helium nuclei (although he did not call them this as he had not yet

ERNEST RUTHERFORD
Genial and avuncular, it was Rutherford's ability to forge and lead a research team, as much as his individual brilliance, that would result in a remarkable string of discoveries in particle physics, and help to transform the way "big science" was done.

© Print Collector | Getty Images

invented the concept of the "nucleus"). In one intriguing experiment he found that a beam of alpha particles fired through air gets slightly scattered, while one passing through a vacuum is not. Repeating the experiment with a thin film of mica (a crystalline mineral), he found more pronounced scattering. As the alpha particles passed through the film, something was causing minor deflections.

These experiments had used photographic film to record the alpha impacts, but Rutherford needed a quantitative approach. Now back in England, at the head of a large team at Manchester University, he worked with Hans Geiger to develop a counting device, where alpha impacts would cause "scintillations," tiny flashes that could be observed and counted through a microscope. In 1908 the device was completed, but the persistent scattering of alpha particles made it difficult to use; Rutherford lamented, "the scattering is the devil."

THE SCATTERING IS THE DEVIL

Rutherford thought he knew what was happening. According to Thomson's prevailing "plum pudding" model of the atom, he noted, "the atom is supposed to consist of a positive sphere of electrification containing an equal quantity of negative electricity in the form of corpuscles." An alternative theory, known as the Saturnine model, because it envisaged a large central mass of positive charge orbited by rings of electrons, had been proposed by the Japanese physicist Hantaro Nagaoka, but Rutherford had been among those to have demolished it by showing that the mathematics were unworkable. Rutherford assumed that his alpha particles were being deflected as they passed between atomic plum puddings, but to find out more he detailed Geiger to make a proper investigation.

© Mrjohncummings | Creative Commons

GEIGER COUNTER
An early Geiger counter, devised by Rutherford's colleague Hans Geiger, as a way to detect and count radioactive emissions. The copper cylinder is partially evacuated and electrically charged, so that each particle of ionizing radiation triggers an electrical discharge.

Geiger experimented with beams of alpha particles passed through foils of gold and other metals, using his scintillating microscope to detect and measure the dispersion of the beams. In late 1908 he took on a bright undergraduate, Ernest Marsden, to help his research, and together they looked for mathematical relationships between alpha dispersion and the thickness of foil (which equated to the number of atoms traversed). Almost casually, Rutherford suggested to Marsden, "See if you can get some effect of alpha-particles directly reflected from a metal surface." Such direct reflection ought not to be compatible with the plum pudding model, and Marsden suspected that Rutherford did not seriously expect him to find anything, later recalling "… it was one of those 'hunches' that perhaps some effect might be observed… Rutherford was ever ready to meet the unexpected and exploit it, where favorable."

To look for this reflection, Geiger and Marsden used a tube of radium to fire a beam of alpha particles at a gold foil. The beam was screened from the scintillating screen so that there could be no direct impacts; only if particles reflected back off the foil could they reach the screen and cause a flash. Much to their surprise, Geiger and Marsden did indeed find some direct reflection, writing in their landmark 1909 paper "On a Diffuse Reflection of the α-Particles," "it seems surprising that some of the α-particles, as the experiment shows, can be turned within a [very thin] layer… of gold through an angle of 90°, and even more. To produce a similar effect by a magnetic field, [an] enormous field… would be required."

Rutherford's account of learning about these results is a classic eureka moment in the history of science. He recalled Geiger coming to him "in great excitement," and telling him "We have been able to get some of the α-particles coming backward… It was quite the most incredible event that has ever happened to me in my life. It was almost as incredible as if you fired a 15-inch shell at a piece of tissue paper and it came back and hit you."

INVENTING THE NUCLEUS

The Geiger-Marsden experiment did not prove quite the turning point that might have been expected from Rutherford's "incredible event" reaction. It took him another year or more to process the result and get past his allegiance to the plum pudding model. But there was no way around the mathematics. In a plum pudding atom the electrical charges are diffused across the entire volume of the atom, making them relatively weak at any point, so that an alpha particle would be deflected only slightly. To account for the massive deflections seen in the experiment, an alpha particle would have to encounter a whole series of gold atoms, building up a cumulative deflection, all in one direction. Such a multiple collision explanation was implausible. "On consideration, I realized that this scattering backward must be the result of a single collision," Rutherford explained, "and when I made calculations I saw that it was impossible to get anything of that order of magnitude [of deflection] unless you took a system in which the greater part of the mass of the atom was concentrated in a minute nucleus. It was then that I had the idea of an atom with a minute massive center, carrying a charge"

"One day [in late 1910/early 1911]," Geiger recalled, "Rutherford, obviously in the best of spirits, came into my room and told me that he now knew what the atom looked like." In May 1911, Rutherford presented his new theory in a paper in the *Philosophical Magazine*, in which he asserted, "the atom consists of a central charge supposed concentrated at a point, and that the large single deflexions of the α and β particles are mainly due to their passage through the strong central field." This was a radical new image of the atom, with electrons scattered at a relatively great distance from a nucleus that has a radius about ten thousand times smaller than the radius of the whole atom, measuring only around ten femtometers, or one hundred thousand billionth of a meter. In other words, the vast majority of the atom is empty space, and the reason that most alpha particles do not scatter is that they simply pass through the voids between nuclei.

Rutherford's triumphant new model started him on a new path of discovery. Under his tutelage, researchers would go on to discover the proton, the neutron, and the positron, knock particles out of the nucleus, and achieve the alchemists' dream of transmutation of the elements.

GRANDFATHER OF CERN
Rutherford's laboratory at the Cavendish research institute at Cambridge University, forerunner of today's immense particle physics laboratories, such as the LHC at CERN.

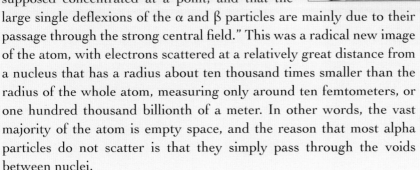

CATEGORY

Astronomy

Biology

Chemistry

Exploration

Mathematics

Medicine

Physics

Technology

GRAVITATIONAL LENSING AND RELATIVITY: EINSTEIN'S BIG IDEAS
1919

Discoverer: Arthur Eddington, Frank Watson Dyson, and Albert Einstein

Circumstances: The obscure theories of a German scientist are put to the test by a team of British astronomers sent to observe an eclipse from distant corners of the world

Consequences: Newton's theories are superseded by Einstein's relativity

... a new Moses come down from the mountain to bring the law, a new Joshua controlling the motion of heavenly bodies. He speaks in strange tongues but wise men aver that the stars testify to his veracity...

Abraham Pais, *Subtle is the Lord: The Science and Life of Albert Einstein* (1982)

A scientific expedition to exotic corners of the world; a total eclipse of the Sun; warps in the fabric of space and time, reordering the heavens; the overthrow of the greatest scientific titan by an obscure upstart, peddling a revolutionary and paradoxical conception of reality. The tale of the 1919 British expedition to observe the eclipse has all the ingredients of a scientific legend; its outcome would make Einstein the most famous scientist in history.

BENDING LIGHT

Try for yourself a thought experiment. Imagine you are traveling in a silent rocketship, which is accelerating through space at a rate of 9.8 m/s^2 (the same rate at which gravity accelerates objects at the surface of the Earth). The rocketship is windowless, but for a tiny porthole over your shoulder, through which a laser beam is shone by an astronaut floating, motionless, in space. From the astronaut's point of view the laser beam travels in a straight line. In the time it takes for the laser beam to travel from the porthole to hit the far side of the cabin, however, the rocketship has moved forward, so the beam does not hit the wall directly opposite the porthole, but a little below that point. If the cabin were filled with smoke you could see the laser beam apparently curving through space. What makes this experience particularly strange is that you have no idea you are in a spaceship, but believe yourself to be in a windowless room back on Earth.

This thought experiment illustrates what is known as the principle of equivalence, one of the tenets of Einstein's theory of relativity, which states that acceleration and gravity are equivalent. In 1911 Einstein predicted, on the basis of this principle, that light will fall in a gravitational field, so that a beam of light from a star, passing close to the Sun, will be deflected from its path. To an observer on Earth, therefore, the star will appear slightly displaced from its actual position; 0.87 arcseconds, according to his calculations.

ALBERT EINSTEIN
The iconic Einstein, his wild corona of white hair and German jacket lending a touch of exoticism. Confirmation of his theory of relativity would make him a global superstar.

In fact this is precisely the degree of displacement predicted by classical Newtonian mechanics, because Newtonian theory also predicts that gravity will act on the particles in a beam of light. A similar value had been published in 1804 by the German astronomer Johann Georg von Soldner, following similar suggestions made by Henry Cavendish in 1784. Shortly after arriving at this Newtonian value, however, Einstein's thinking about relativity began to evolve as he assimilated the theories of the German physicist Hermann Minkowski (1864–1909), who had proclaimed in 1908 that "Henceforth space by itself, and time by itself, are doomed to fade away into mere shadows, and only a kind of union of the two will preserve an independent reality."

Einstein realized that space and time are stitched together into a fabric, which is warped by the presence of matter, and that gravity is the consequence of this warping of spacetime. Light would still travel the shortest distance between two points, but on a curved surface this is not a straight line but a geodesic; thus the hypothetical beam of starlight passing close to the Sun would be additionally deflected by the curvature of spacetime. In 1916 Einstein calculated the total deflection at the limb of the Sun would be twice the Newtonian value, or about 1.75 arcseconds (an arcsecond is 1/3600 of a degree, so this is 0.0004861°). This is the angle made by a right triangle 1 inch (2.5 centimeters) high and 1.9 miles (3.1 kilometers) long.

Although minute, such a deviation might be observable, providing a dramatic test of Einstein's radical new theory. Einstein had written to leading astronomers as early as 1913, trying to interest them in such a test, but it was not until 1917 that anyone took up his offer (perhaps serendipitously, given that it was only in 1916 that he made the correct calculation). Partly this was due to the small matter of a world war.

THE ECLIPSE OF 1919

A revolutionary theory by a German scientist, which threatened to overthrow the secular god of British science, was not well received in Britain during World War I. Yet Einstein had one major proponent, the British astronomer Arthur Eddington. He made sure that Einstein's papers were translated into English, and popularized the confirmation of relativity suggested by the perihelion advance of Mercury. This was a longstanding astronomical conundrum concerning the way that the perihelion (closest approach to the Sun) of Mercury shifted, which could not be explained by classical Newtonian mechanics, but which was perfectly accounted for by relativity. Einstein experienced heart palpitations on learning of this confirmation, writing to a colleague that he was "for several days beside myself with joyous excitement."

Among those alerted to Einstein's relativity, although he was skeptical of it, was the Astronomer Royal Sir Frank Watson Dyson. In 1917 he realized that the perfect opportunity to test the light deflection prediction was coming up. Observations of the position of stars close to the Sun faced one obvious drawback: stars are not visible when the Sun is shining. But during a total eclipse of the Sun, bright stars do become visible, even quite close to the limb. Dyson knew that at the upcoming total eclipse of May 1919, the Sun would be in the midst of a cluster of bright stars called the Hyades. He put plans in motion; as chairman of the Joint Permanent Eclipse Committee of the Royal Society and the Royal Astronomical Society, he put Eddington in charge of preparations for an expedition to observe the 1919 eclipse.

TOTAL ECLIPSE
One of the Sobral photographs of the total eclipse of 1919. Eddington wrote that "applications for these [prints] from astronomers, who wish to assure themselves of the quality of the photographs, will be considered."

In March 1919, Eddington traveled to the island of Principe just off the coast of western Africa, while two of Dyson's assistants, Andrew Crommelin and Charles Davidson, took up position at Sobral in northern Brazil. On May 29th, despite some problems with cloud cover and malfunctioning cameras, both teams were able to take photographs of the Hyades stars around the Sun during the eclipse. Eddington had taken "control" photographs of the cluster before leaving, but had to wait for the Sobral team to take their "control" pictures a couple of months later (by which time the Sun was rising in a different part of the sky, giving an undeflected view of the stars). By September both sets of results were being analyzed. There have subsequently been controversies over the data analysis, and allegations of bias by Eddington in particular (Dyson famously suggested that if the value obtained was not the Einstein predicted one, "Eddington will go mad"). But it is generally accepted that the results confirmed Einstein's theory, matching his predicted values with an accuracy of about 30 percent.

Dyson's reaction was cautious. In a letter of March 1920, to Frank Schlesinger, director of the Yale Observatory, he admitted, "The result was contrary to my expectations, but since we obtained it I have tried to understand the Relativity business, & it is certainly very comprehensive, though elusive and difficult." The popular press had a very different response. When the results of the experiment were announced in November 1919, they generated extraordinary headlines. "Revolution in Science," heralded the *Times* of London, on November 7th, "New

MORE OR LESS AGOG

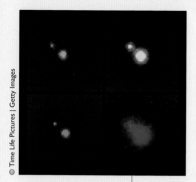

LENSING IN ACTION
These images from the
Hubble space telescope
show a single quasar that
presents a double image
because its light has been
bent round either side
of an intervening mass
such as a galaxy.

Theory of the Universe: Newton's Ideas Overthrown."
The front page of the *New York Times* for November 10th
trumpeted: "Lights all askew in the heavens: Men of science
more or less agog over results of eclipse observations.
Einstein theory triumphs. Stars Not Where They Seemed or
Were Calculated to be, but Nobody Need Worry."

Before this Einstein had been a complete unknown outside of
scientific circles. Now he became a global celebrity. He won
the Nobel Prize and in March 1921 was given a ticker-tape
parade down Broadway in New York City—the only scientist to ever
receive such an honor. There can be little doubt that relativity was a
revolutionary development in the history of science, but how to account
for the extraordinary public enthusiasm for Einstein the man? The
context of the discovery must have been a major factor; the expedition
itself may have been undertaken—particularly by Eddington—in the
spirit of reconciliation, with British scientists confirming a "German"
theory. Thus in the aftermath of the Great War came portents of a new
conception of reality, heralded by intellectual cooperation triumphing
over petty nationalism. Einstein's great biographer, the Dutch-American
physicist Abraham Pais, argues that there was a widespread appetite to
see Einstein as an almost Messianic figure:

> A new man appears abruptly… He carries the message of a new order in the
> universe… He fulfills two profound needs in man, the need to know and the need
> not to know but to believe.

Einstein himself apparently never had any doubt about what the eclipse
expedition would reveal. Asked how he would have felt if reality had
not conformed to his theory, he replied, "Then I would feel sorry for the
good Lord. The theory is correct!"

PENICILLIN: THE INVENTION OF ANTIBIOTICS

1928

Discoverer: Alexander Fleming

Circumstances: Fleming forgets to put away a dish of bacterial culture when he goes on vacation. By chance the dish has been contaminated with a mold spore

Consequences: Discovers antibiotic effects of penicillin

If I may offer advice to the young laboratory worker, it would be this—never neglect an extraordinary appearance or happening.

Alexander Fleming

The quintessential accidental revelation in the history of medicine, Sir Alexander Fleming's discovery of the antibiotic penicillin ranks as one of the foremost myths in the history of science, alongside Newton's apple and Archimedes's overflowing bathtub. Yet while the discovery was indeed fortuitous and momentous—enough to earn Fleming a Nobel prize—it did not come out of a vacuum, nor did it arrive as a fully formed medical breakthrough.

MOLD IN ACTION

Fleming's discovery came when he noticed that a patch of mold contaminating one of his bacterial culture dishes had a bacteria-free ring around it; he subsequently characterized the mold as a *Penicillium* species, and hence named the antibacterial substance it produced as penicillin. But this was by no means the first inkling of the antibacterial powers of mold, which had been observed several times before. Ancient Egyptian physicians applied poultices of moldy bread to infected wounds, suggesting that they probably recognized the phenomenon, while the *Penicillium* mold itself had been linked with an absence of bacterial contamination in 1874, by William Roberts. Three years after this Louis Pasteur and Jules Francois Joubert observed that growth of anthrax bacilli could be inhibited by molds, and in 1897 a French researcher, Ernest Duchesne, showed by experiment that a *Penicillium* mold could both eliminate bacterial growth in culture (i.e. when it was grown on a nutrient jelly in a glass dish), and could apparently stop the development of typhoid in animals infected with normally lethal typhoid bacteria. Duchesne urged further research on his findings but tragically died of tuberculosis—a disease now treated with antibiotics—before he could advance his discovery.

ALEXANDER FLEMING

Fleming shown in his laboratory at St. Mary's in London where he discovered penicillin. He is in the process of inoculating the agar jelly in a petri dish, the primary tool of microbiology.

In other words, Alexander Fleming was not the first to make the link between this specific species of mold and the antibacterial phenomenon; as he reflected during his 1945 Nobel acceptance lecture, "To my generation of bacteriologists the inhibition of one microbe by another was commonplace." He was, however, perhaps uniquely placed to appreciate that the phenomenon was important and worth further investigation, having in 1921 already discovered and isolated one of the first antibiotic agents from his own saliva, a discovery in which chance occurrence had also played a part.

At the time of his discovery, Fleming was Professor of Bacteriology at St. Mary's Hospital in London, where he had spent most of his career, starting as a research scientist under the auspices of Sir Almroth Wright, a pioneer in vaccine therapy who strongly believed in the primacy of the body's own defenses over external agents (such as the use of antiseptics, which had become medical doctrine in the late 19th century). Fleming took his research direction from Wright, notably while working at a wound research laboratory in France during World War I, where he discovered that ghastly infections like gangrene were better treated by letting the body's natural reactions—such as pus—do their work, while antiseptics actually favored the growth of the disease-causing microbes.

Back at St. Mary's after the war, Fleming discovered a powerful antibiotic agent in his own snot, after testing his nasal secretions while he had a bad cold. Something in the human body's own secretions, including saliva, tears, and milk, had the power to break apart many species of microbe, a phenomenon known as lysis. Fleming isolated the agent and named it lysozyme, but it appeared only to be effective against mostly harmless airborne bacteria rather than pathogenic (disease-causing) germs. It also proved hard to collect and concentrate lysozyme into more potent formulations, foreshadowing the problems that would impede research on penicillin. But Fleming developed techniques for isolating and testing substances that would later serve him well.

In 1928 Fleming was doing research on *Staphylococcus* bacteria, a common component of human bacterial flora that is normally harmless but which can, especially in patients with a weakened immune system, cause terrible boils and other infections. In August he left London for a two-week vacation, and on his return on September 3rd, 1928 he began to go through the petri dishes he had set up before he left. A petri dish is a small, circular, shallow glass dish used in microbial research. Each dish is filled with clear agar jelly, a nutrient-rich jelly on which microbes such as bacteria and fungi will thrive, forming colonies that grow as spots or stripes (depending on how the jelly is inoculated with the microbe in the first place). Each dish must be prepared under careful, contamination-free conditions, so that only the desired test organisms will form colonies, or else stray microbes from the environment will contaminate the dish and ruin the experiment.

AN EXTRAORDINARY APPEARANCE

This is exactly what happened to the fateful dish prepared for Fleming in mid-August, which was evidently contaminated by a mold spore, commonly held to have floated in through the laboratory window. In practice the lab windows would have been kept shut for precisely this reason, and the mold spore possibly came up the stairwell from a lab on a lower floor where molds were being researched. To add another layer of serendipity, the dish in question was not placed as intended in the incubator, which keeps dishes at the optimal temperature for bacterial growth but which would have prevented the mold spore from thriving; instead by chance the bench-top environment provided just the right temperature for both bacteria and mold to grow.

MOLD MEMENTO
A sample of penicillium mold that Fleming gave to his colleague at St. Mary's, Douglas Macleod, and which was later mounted and bound in a case by Macleod's wife.

When Fleming looked at the dish that had accidentally been left out of the incubator, he saw a spot of mold amidst the staphylococcus colonies, and noted that around the mold spot there was a zone of clear jelly where no colonies were growing. "This was an extraordinary appearance and seemed to demand investigation," Fleming told an audience at his Nobel lecture in 1945, "so the mold was isolated in pure culture and some of its properties were determined."

Fleming managed to isolate the active ingredient from the mold; he called it penicillin. But even though he noted its bacteria-killing properties, it does not seem to have occurred to him to test its therapeutic properties in an animal subject. Fleming apparently conceived only that penicillin might prove useful in isolating certain strains of bacteria, and perhaps as a topical antibacterial. He published his results to little fanfare in the *British Journal of Experimental Pathology* in June 1929.

THE EXTRACTION IS MURDER

Penicillin only became a legendary discovery thanks to the efforts of the men who would share the Nobel Prize with Fleming, Howard Florey and Ernst Chain from the Sir William Dunn School of Pathology at Oxford University, along with many others. In 1939 Chain worked out a laborious method to purify and concentrate penicillin. It involved growing the mold in a nutrient-rich broth, scooping it off the surface and repeatedly freeze-drying its juice. It took many dozens of pints of broth to produce even enough penicillin to cover a fingernail.

In 1940 Florey made the vital demonstration that penicillin could indeed "cure" a common illness, infecting eight mice with deadly hemolytic streptococci, a bacteria responsible for many postpartum deaths in new mothers, and giving half of them penicillin. There was great excitement when the four treated mice survived, and the publication of the findings in *The Lancet* in August 1940 caused a sensation in the field. On February 12th, 1941, a 43-year-old policeman, Albert Alexander, became the first human to be treated with penicillin. His life-threatening abscesses went into remission, but after a few days the meager supplies of penicillin ran out and he died.

PENICILLIN PRODUCTION
A photo from 1943 shows British penicillin production, still relatively small-scale compared to the increasingly industrialized process in the US, where penicillin would become available over the counter by 1945.

The race was now on to scale up production and make penicillin treatment a reality for a world at war. At Oxford a team of "penicillin girls" had been employed to tend the mold broths, with all manner of vessels pressed into service, from baths to bedpans. But this was clearly inadequate, and the Allied authorities identified penicillin as a project that would benefit from industrial levels of research and investment. The difficulties facing the scale-up effort were encapsulated by Pfizer's John L. Smith: "The mold is as temperamental as an opera singer, the yields are low, the isolation is difficult, the extraction is murder, the purification invites disaster, and the assay is unsatisfactory." But with the might and money of American pharmaceutical companies in harness to produce a sort of medicinal Manhattan Project, production accelerated incredibly. In the US, production of the drug rocketed from 21 billion units in 1943, to 1,663 billion units in 1944, to more than 6.8 trillion units in 1945, with the process moving from one-liter flasks with 1 percent yield to 10,000 gallon tanks at 80–90 percent yield. By March 1945 the Americans were able make penicillin freely available over the counter.

In his 1945 Nobel lecture, Fleming related his impressions on visiting the penicillin "factories" in America: "To me it was of especial interest to see how a simple observation made in a hospital bacteriological laboratory in London had eventually developed into a large industry and how what everyone at one time thought was merely one of my toys had by purification become the nearest approach to the ideal substance for curing many of our common infections."

NUCLEAR FISSION: SPLITTING THE ATOM

1938

Discoverer: Lise Meitner, Otto Frisch, Otto Hahn, and Fritz Strassmann

Circumstances: Seeking to create heavier elements by firing neutrons at uranium, scientists are puzzled to find lighter elements instead

Consequences: Nuclear fission and the chain reaction; new energy sources

Listen, young man, I want to tell you about something that's very important, that's recently happened in physics… Let me tell you about fission.

Niels Bohr speaking to Herbert Anderson at Columbia University, 1939

The discovery of nuclear fission is a thrilling and breathless tale of the interplay of ideas between great minds in a world on the brink of war, with the shadow of awful possibility always looming. Ascribing the discovery to a single person or team is difficult, because it required a synthesis of experimental evidence and theoretical insight.

The discovery of radioactivity had raised the prospect of obtaining a new and wonderful source of power in the atom. Reading about the discovery of radioactive decay, H. G. Wells was inspired to create fictional atomic bombs in his 1913 novel *The World Set Free*, which in turn would inspire the next generation of nuclear scientists to consider the real thing.

PURELY SCIENTIFIC

In 1932, working under Rutherford at the Cavendish in Cambridge, John Cockcroft and Ernest Walton became the first people to "split the atom," using a new particle accelerator to fire protons at lithium atoms, achieving transmutation of the elements and liberating some energy. The *New York Times* enthused that "Science has obtained conclusive proof from recent experiments that the innermost citadel of matter, the nucleus of the atom, can be smashed, yielding tremendous amounts of energy." But even now most scientists agreed with remarks made by Rutherford in 1933 and published in *Nature*: "we cannot control atomic energy to an extent which would be of any value commercially, and I believe we are not likely ever to be able to do so… Our interest in the matter is purely scientific."

A notable dissenter was the Hungarian physicist Leo Szilard. Inspired by Wells's novel, he was again fired by the news that, also in 1932, James Chadwick at the Cavendish had discovered proof of Rutherford's long predicted neutron—a subatomic particle that makes up atomic nuclei along with protons, but which, unlike protons, has no charge and thus can more easily approach and become absorbed by other atomic nuclei. Szilard realized that if atoms of a radioactive element like uranium could be made to release enough neutrons, these neutrons in turn would trigger further neutron release by other atoms, setting off a chain reaction that could liberate vast amounts of energy. Since it was not clear how such a neutron release could be triggered, there was little interest in Szilard's idea.

STRANGE RESULTS

Research teams all across Europe and America now started to bombard heavy elements such as uranium with particles such as protons and neutrons. It was assumed that uranium nuclei would capture the particles and be transmuted into heavier elements, known as transuranics because they were beyond uranium in the Periodic Table. In Germany Otto Hahn, Lise Meitner, and Fritz Strassmann were researching what happened when uranium was bombarded with neutrons, but Nazi persecution forced the Jewish Meitner to flee to Sweden in July 1938. Hahn recalled that "Strassmann and myself, we had to work alone again and in the fall of '38 we found strange results…" Looking at the products of bombardment of uranium, they found no evidence of transuranics—only of lighter elements: "we could be absolutely sure that there could be nothing else but either radium or barium."

HAPPIER DAYS
Otto Hahn and Lise Meitner working together in the laboratory in 1913. Twenty-five years later Nazi persecution would force her out of German science.

This seemed impossible—a single neutron might knock three or four protons off a heavy atom and produce radium, but how could it smash off the 100 or more necessary to produce light elements such as barium? Since radium and barium are very hard to distinguish chemically, Hahn said, "we could conclude that the substances could be really only radium… because barium was prohibited by the physicists that we didn't dare to think it barium in those times. We always tried to explain what is wrong in our experiments… we poor chemists… [we were] so afraid of these physics people."

In late December Meitner was visited in Sweden by her nephew, the physicist Otto Frisch. He found her "brooding over a letter of Hahn," reporting the "strange results." They discussed it at great length: "Barium, I don't believe it. There's some mistake… It's fantastic. It's quite impossible…" Finally Meitner and Hahn drew similar sketches of what might actually be going on, showing a nucleus contracting around the middle, pinching in half. "Couldn't it be this sort of thing?" wondered Meitner.

Meitner and her nephew had just come up with the concept of nuclear fission, as it came to be called for its resemblance to the process of cellular fission such as when an amoeba divides into two smaller daughter cells. Frisch recalled that the term fission "was suggested to me by an American biologist, William A. Arnold, whom I asked what you call it when a cell divides itself."

In nuclear fission, a large nucleus like that of uranium captures a neutron and becomes unstable, splitting into smaller nuclei (such as barium nuclei). M and F calculated the energy associated with the splitting apart of the nuclear fragments to be around 200 million electron volts (MeV) per uranium atom. By comparison, the most energetic chemical reactions release approximately 5 eV per atom. Where could this energy come from? The experimental data showed that the products of fission—the fragments left after fission—did not add up to the total mass of the uranium nucleus before fission. A tiny amount of mass had disappeared, and using Einstein's formula $e=mc^2$, they calculated that it equated to... 200 MeV of energy.

Arriving back in Copenhagen, Frisch caught Bohr "on the point of leaving for America, and I just managed to catch him for five minutes and tell him what we had done. And I hadn't spoken for half a minute when he struck his head with his fist and said, 'Oh, what idiots we have been that we haven't seen that before. Of course this is exactly as it must be... This is very beautiful.'" Frisch promptly set to work looking for experimental evidence that would support the theory. He put together an apparatus with an ionization chamber connected to an oscilloscope, which would display pulses caused by the passage of particles, reasoning that particles produced by nuclear fission would be the most energetic yet seen and thus would produce large pulses: "By three in the morning I had the evidence of the big pulses."

THIS IS VERY BEAUTIFUL

The news about fission sparked renewed interest in nuclear power. "The possibility of harnessing the energy of the atom crops up again," observed a *New York Times* editorial in February 1939. Scientists modeling the process realized that fission of uranium might indeed release "free" or "secondary" neutrons. "If there were enough of these," speculated the American physicist John Dunning, "then the long-sought-for key to a self-sustaining nuclear energy release might indeed be here. World War II was pretty clearly in the offing at that time and all of us recognized the far-reaching consequences that might be possible if fission could really be developed."

FATAL FISSION
The mushroom cloud over Nagasaki on August 9th, 1945. At the base of the cloud the explosion of the "Fat Man" plutonium bomb has already killed around 35,000 people; in the coming months the same number again will die of their injuries and exposure to radiation.

Szilard had long since been urging scientists to look into this possibility. Now there was a race to find out whether uranium fission did indeed produce "free" neutrons, and soon teams in Paris and New York both confirmed that it did: a chain reaction would be possible. A controlled chain reaction, where the cascade of released neutrons can be slowed and damped with control rods, would produce manageable heating, which in turn could be the basis for power generation. Three years later Enrico Fermi's team switched on the first nuclear pile, which they had constructed beneath the bleachers of the football stadium at the University of Chicago. Arthur Holly Compton recalled the scene:

Fermi gave the order to withdraw the control rod another foot. We knew that that was going to be the real test. The geiger counters registering the neutrons from the reactor began to click faster and faster till their sound became a rattle. The reaction grew until there might be danger from the radiation up on the platform where we were standing. "Throw in the safety rods," came Fermi's order. The rattle of the counters fell to a slow series of clicks. For the first time, atomic power had been released. It had been controlled and stopped. Somebody handed Fermi a bottle of Italian wine and a little cheer went up.

Leo Szilard had a rather darker reaction: "I shook hands with Fermi and I said that I thought this day would go down as a black day in the history of mankind. I was quite aware of the dangers… something had to be done if the Germans get the bomb before we have it… [they] would have forced us to surrender if we didn't have bombs also. We had no choice…" The race to create a fission bomb was on.

THE LASCAUX CAVES: A NEW WORLD OF PREHISTORIC ART

1940

CATEGORY

Astronomy

Biology

Chemistry

Exploration

Mathematics

Medicine

Physics

Technology

Discoverer: Marcel Ravidat, Robot the dog, and friends

Circumstances: Hunting for a legendary treasure tunnel, four friends and a dog stumble upon a mysterious hole in the ground

Consequences: Discovery of a treasure trove of paleolithic art

We have invented nothing.

Pablo Picasso's reaction on visiting Lascaux

© Ralph Morse | Getty Images

The discovery of the Lascaux cave paintings in 1940 reads like something from a children's novel, with a gang of kids—and their dog—crawling down a dark hole to find themselves staring in wonder at long extinct animals drawn by our distant ancestors. The amazing art they revealed changed the conception of paleolithic humanity forever.

DOWN THE RABBIT HOLE

The precise sequence of events differs with source, but accounts agree that on September 12th, 1940, four teenagers and their dog were looking for adventure in the Dordogne region of France. In some versions one of the boys, Marcel Ravidat, had already spotted a promising foxes' earth some days earlier and had brought back help to investigate, while in others their dog Robot discovers the hole beneath an uprooted tree. However it happened, the teenagers were excited because they hoped the mysterious hole in the ground might be related to a local legend, about a tunnel in the Montignac woods linking an old castle to the Manor of Lascaux, and which might lead to a hidden hoard of treasure.

Widening the narrow hole with their penknives, the teenagers lowered themselves into the unknown, and slid down a 49-foot (15-meter) shaft into an underground chamber. Once inside they used their oil lantern to look around. From out of the gloom emerged a pure white frieze covered with "a cavalcade of animals larger than life," recalled the youngest of the boys, 14-year-old Jacques Marsal; "each animal seemed to be moving." They were standing in what is now called the Hall of Bulls, sometimes described as the Sistine Chapel of paleolithic art, a 65-foot (20-meter) long chamber decorated with exquisite depictions of animals, including 17 horses, 11 cows and bulls, and 6 stags. The ancient artists who painted them picked a perfect natural canvas: a strip of wall covered with pure white calcite (crystallized chalk).

The boys kept their discovery secret for a day or two, returning to explore further and even making some money by charging admission to villagers, but they soon informed the local schoolmaster, Leon Laval, about the find. Once he had overcome his suspicion that the boys were playing a prank on him, Laval too lowered himself down into the cave; a member of the prehistoric society of Montignac, he immediately recognized that the art was probably paleolithic. By chance, the Abbé Henri Breuil, one of the leading authorities on prehistory, was in the region and learned of the discovery. On September 21st he visited for himself, confirming the authenticity and antiquity of the find. Now the professionals moved in, and the work of cataloging and examining the art and exploring the archaeology of the site commenced.

Lascaux was probably the finest collection of paleolithic art yet discovered. It swept away 19th-century notions about primitive cavemen and simple savages, proving that our Stone Age ancestors possessed sophisticated aesthetic capabilities and great technical skill. The Lascaux images (there are engravings as well as paintings, and in fact the former outnumber the latter although they are not as spectacular) exhibit amazing control of color, and figurative and abstract representation. The painting called "the Crossed Bison," in the chamber known as the Nave, has crossed hind legs to indicate perspective.

OLD MASTERS

The art at Lascaux includes nearly 2,000 figures, in three main categories: animals, human figures (which are very sparse), and abstract markings. The most striking images are in the Great Hall of the Bulls, which includes four enormous black bulls or aurochs (giant wild cattle now extinct). The Great Bull, at 17 feet (5.2 m) in length, is the largest animal image in cave art. The most numerous animals are horses, but there are also stags, felines, a bear, and a rhinoceros. There is one image of a man, in the Shaft of the Dead Man, where he features in a striking tableaux falling away from an apparently injured or dead aurochs.

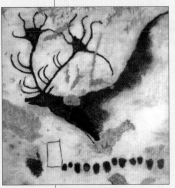

GIANT ELK
Replica of one of the remarkable artworks from Lascaux, known as the Black Stag. It is probably a depiction of a Megaloceros, known as the Giant or Irish Elk. Note the geometric shapes below the beast.

Lascaux and other examples of so-called parietal art from the Upper Paleolithic (the last phase of the "old" Stone Age, from around 50,000 to 10,000 years ago) in Europe were so startling and advanced that they formed the basis of a theory about the evolution of human cognition (thinking ability), the so-called Great Leap or "creative explosion." This theory claims that around 40,000 years ago prehistoric humans underwent a step change in cognitive and thus creative ability, demonstrated by a sudden flowering of arts and crafts such as Venus figurines and parietal art. The cause was thought to be a genetic mutation causing a crucial change in the prehistoric human brain.

In part this theory depended on the dating of the parietal and other arts. Lascaux has been dated to the Solutrean/Lower Magdalenian periods, around 18,000 years ago, relatively late in the Upper Paleolithic, and it was thought that it represented a natural progression in an artistic tradition that had begun around 40,000 years ago, having started from essentially nothing. In fact subsequent findings have shattered this conceit. Lascaux's main rival as the most exquisite example of prehistoric parietal art, the Chauvet cave, astonished the scientific community when direct radiocarbon dating revealed it to date to 40,000 years ago. In other words, the supposedly evolving tradition was already fully developed at least as early as the start of the Upper Paleolithic, with some younger sites displaying much less highly developed aesthetics. In the 21st century, a host of finds from Africa has pushed back the date of earliest evidence of human artistic creativity to 100,000 years ago and more. It seems likely that environmental, geographical, and cultural factors have combined to increase the evidence available for European Upper Paleolithic art relative to older, non-European art. In other words, there probably was no Great Leap 40,000 years ago, but a much older and more heterogeneous and fluctuating tradition stretching back to human origins in Africa.

Whether or not they signal an evolutionary leap in human cognition, the Lascaux artworks have great power and, presumably, once held great meaning. It is generally assumed that parietal art was magical in nature, and may have been created by shamans or equivalent figures. The most symbolically significant art is often located in the farthest and least accessible portions of cave systems, indicating that it had a special, esoteric role. Current thinking emphasises the likelihood that the cave walls were seen as a kind of membrane between worlds, with the paintings and engravings thus offering direct communion with spirit creatures or animal manifestations of humans undergoing mystical experiences. The abstract markings in Lascaux and other sites have been linked with entoptic phenomena: patterns and lights seen in the dark or when sensorily deprived, or in altered states of consciousness. These are visual perceptions created by internal brain activity in the absence of external stimuli, and it is thought likely that the artists, either in the dark or in ecstatic trances or under the influence of hallucinogens—or a combination of all three—directly copied their entoptic visions onto the rocks.

THE DOUBLE HELIX: THE MYSTERY OF DNA IS DECODED

1953

CATEGORY

Astronomy

Biology

Chemistry

Exploration

Mathematics

Medicine

Physics

Technology

Discoverer: Francis Crick and James Watson

Circumstances: With three teams racing to solve the mystery of the structure of DNA, Crick and Watson are given unauthorized access to a key piece of evidence found by a rival

Consequences: The structure of DNA; the secret code of life

We seem to have made a big discovery.

Francis Crick to his wife, the evening of February 28th, 1953

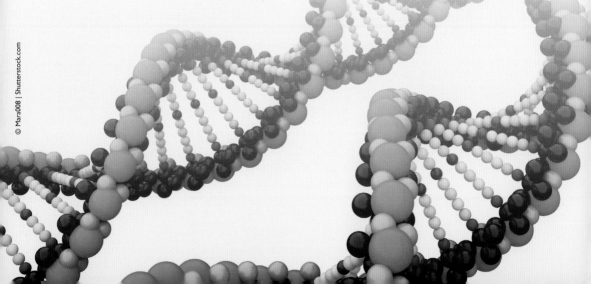

The discovery of the structure of DNA is perhaps the greatest landmark in the history of the life sciences, revolutionizing or creating fields from genetic engineering to agriculture, crime fighting to evolutionary biology. It revealed to humanity the language of life and let us read the Book of Nature as never before.

DARK CONTOURS

The discovery by Francis Crick and John Watson of the structure of the DNA molecule was the culmination of a long series of research that began slowly but gathered pace to reach a thrilling climax. The quest to identify the agent of heredity—the way in which heritable characteristics are passed from parent to offspring—dated back to Darwin and beyond, although it turned out that it had been hiding in plain sight for decades. Deoxyribonucleic acid (DNA) was first identified in 1869, by the Swiss biochemist Friedrich Miescher. In the late 1830s, with the rise of Cell Theory, the cell had been identified as the basic unit of organisms. Miescher was investigating the chemical makeup of different parts of the cell, and was particularly interested in the nucleus. It was known that white blood cells have especially large cell nuclei, and a good source of white blood cells is pus, from wounds. Accordingly Miescher would stop by a clinic every morning on the way to work to pick up pus-soaked bandages. In the lab he mixed the cells with an alkali to make the nuclei burst open, and was able to extract from the contents thus released a hitherto unknown organic molecule, which he called nuclein.

FRANCIS CRICK
After discovering the genetic code Crick went on to work on the molecular biology of gene expression.

Nuclein proved to be a phosphorous containing acid unlike better known organic chemicals such as carbohydrates and proteins. One of the most remarkable properties of this nucleic acid, as it came to be known, is its ability to form incredibly long chains, but Miescher did not realize this as he must have been working with relatively tiny fragments, a problem that would mislead researchers into the 20th century.

In 1879 the German biologist Walther Flemming discovered chromosomes—tiny filaments in the nucleus, which he initially called chromatin because of its susceptibility to staining. Watching cell division through the microscope, biologists could see the stained chromosomes duplicating and dividing so that each daughter cell would carry a full set. Evidently these structures played a key role in heredity. Analysis showed that chromosomes contained both DNA and proteins, and in 1884 Oskar Hertwig claimed, "I believe that I have at least made it highly probable that nuclein is the substance that is responsible… for the transmission of hereditary characteristics." But most biologists disagreed; proteins were revealing themselves to be huge molecules of great complexity, capable of carrying out a great range of biochemical tasks in the cell— in other words, much more suited to the task of carrying the blueprint for an organism.

JAMES WATSON
Watson has written two of the most influential books in biology: *The Double Helix*, his account of the discovery of the structure of DNA, and *Molecular Biology of the Gene*, perhaps the standard text book in molecular genetics.

The chemistry of DNA was further elucidated in 1900, when it was found that nuclein consisted of three components: a phosphate, a sugar called deoxyribose, and four bases called adenine, cytosine, guanine, and thymine, commonly referred to by their initials. Lithuanian-American biochemist Phoebus Levene discovered that these three components are connected in the order phosphate-sugar-base, to give units he called nucleotides, and that the phosphate groups can connect together to give a long DNA molecule consisting of a chain of nucleotides with a phosphate backbone. But Levene thought that DNA molecules would only be around 10 bases long, when in fact it is now known that DNA molecules in nature are millions of bases long. A single human cell contains DNA sequences totalling 3.2 billion bases in length, which would stretch over 3.3 feet (1 meter) if laid end to end. Levene was also responsible for the tetranucleotide hypothesis, which believed the ratio of the four bases to be the same in all DNA everywhere, no matter what the organism. In other words, DNA was seen as a molecule with no particular information content. Proteins, meanwhile, had been shown to be composed of chains of amino acids, of which there are 20 different types; they seemed much more likely to be the medium for any form of genetic code.

In 1928 the English microbiologist Fred Griffith had confirmed what had been long suspected. He found that two strains of bacteria with different appearances and properties could swap properties, so that harmless rough pneumococci bacteria became deadly smooth pneumococci. Griffith proposed that a "transforming principle" must have been passed between strains, and in 1944 American microbiologist

Oswald Avery and his team had identified this principle as DNA. The agent of heredity had been unmasked, but this introduced a new mystery: how could such an unpromising molecule carry the genetic blueprint? The race was on to determine the structure of DNA and solve the riddle. "I saw before me in dark contours the beginning of a grammar of biology," wrote the Austrian chemist Erwin Chargaff. "Avery gave us the first text of a new language, or rather he showed us where to look for it. I resolved to search for this text." Chargaff disproved Levene's tetranucleotide hypothesis, showing that each species had its own unique proportion of bases but that in all cases the ratio of A to T and C to G was one to one. This would prove to be a vital clue.

MODEL SCIENTISTS?

By 1952 there were three teams working to crack the DNA conundrum. The leading team was at King's College, London, where biophysicist Maurice Wilkins had obtained high-quality samples of DNA and was examining them with X-ray crystallography. This was a powerful tool for imaging the structure of molecules, which had led to the elucidation of the structure of proteins such as hemoglobin. When the brilliant X-ray crystallographer Rosalind Franklin joined the team at Kings, however, there was immediate tension, and the antipathy between her and Wilkins retarded their research effort. Meanwhile in America the brilliant biochemist Linus Pauling was also turning his attention to DNA. His novel approach of building models using rods and balls had helped him to scoop British researchers in his discovery of an important helical structure in some proteins. Franklin's X-ray pictures began to suggest that maybe DNA also had a helical structure.

The news that Pauling had entered the race galvanized the third research team, the pairing of English biophysicist Francis Crick and young American microbiologist James Watson, at the Cavendish in Cambridge. They performed no experiments or imaging studies of their own, but intended to use the model-making approach to solve the puzzle, working on the data published by others. They were friendly with Wilkins, and in a fateful meeting he showed them an X-ray image taken by Franklin and her assistant Raymond Gosling, known as Photo 51, which clearly demonstrated the helical nature of the molecule. She herself had ignored it for the best part of eight months, her feud with Wilkins and the others blinding her to its importance, and by the time she started to appreciate its significance, in January 1953, Wilkins had already showed it to Watson. In his famous memoir of the DNA race, *The Double Helix*, Watson recalls that "The instant I saw the picture, my mouth fell open and my heart began to race."

Pauling had published a model of DNA as a helical structure, but he had mistakenly arranged it with the phosphate backbone on the inside and the bases on the outside. Crick and Watson realized that if the bases were on the inside, the base pairings revealed by Chargaff's ratios meant that A bases would form strong bonds with their opposite T bases, and likewise for C and G. Their model revealed that this arrangement fit perfectly with a double helix with one helix slightly out of step with the other, with the phosphate backbones of each strand as the poles of a ladder, and the base pairings as the rungs. They completed their model on February 28th, 1953. Watson recalls that Crick went to the pub and announced that they had "found the secret of life." Crick himself did not recall this, but remembered going home and telling his wife "that we seemed to have made a big discovery," although, "years later she told me that she hadn't believed a word of it. [She told me] "You were always coming home and saying things like that, so naturally I thought nothing of it.""

They published their theory in a one-page letter to *Nature*, proudly announcing, "We wish to put forward a radically different structure for the salt of deoxyribose nucleic acid. This structure has novel features which are of considerable biological interest." The most important aspect of their discovery was only alluded to: "It has not escaped our notice that the specific pairing we have postulated immediately suggests a possible copying mechanism for the genetic material." What they meant, and what is so beautiful and elegant about the structure of DNA, is that form is function. In order to act as the genetic medium, DNA must be able to specify its own duplication, and the base pairing arrangement achieves precisely this. When the DNA helix unzips down the middle and the two sides separate, each half acts as the template for its complement. Crick, Watson, and Wilkins shared the 1962 Nobel Prize for their parts in the discovery. Franklin, tragically, had died of cancer in 1958.

DISCOVERY BY MODEL
Physical model building was a crucial step in Crick and Watson's discovery of the DNA structure. This is a reconstruction of their original model, using some of the same metal plates (representing nucleotide bases) as the original.

© Science & Society Picture Library | Getty Images

CATEGORY

Astronomy

Biology

Chemistry

Exploration

Mathematics

Medicine

Physics

Technology

THE BUTTERFLY EFFECT: EDWARD LORENZ AND THE BIRTH OF CHAOS THEORY

1961

Discoverer: Edward Lorenz

Circumstances: Rerunning a computerized weather simulation without changing the starting conditions, Lorenz is amazed to get a completely different outcome

Consequences: Discovers that the universe is fundamentally indeterminate

The discovery of deterministic chaos has profoundly influenced a wide range of basic sciences and brought about one of the most dramatic changes in mankind's view of nature since Sir Isaac Newton.

Edward Lorenz's citation for the Kyoto Prize of the Inamori Foundation, 1991

Few discoveries can have had a more fundamental impact on a greater range of sciences than chaos theory, a branch of mathematics and a way of looking at nature that radically resets the way science approaches the universe. When Edward Lorenz discovered the butterfly's wing effect, however, he was simply bringing into focus a set of concepts that mathematics had known about for over 70 years.

In 1961 American mathematician and meteorologist Edward Lorenz was running a simulation of the weather using a computer program to crunch the numbers for him. A computer program is the ultimate determinate system, in that it goes through a specified sequence of operations in a specified order the exact same way every time it is run. Putting the same data into the simulation each time must result in getting the same output, because the exact same operations are carried out. So when Lorenz ran the same simulation twice but got completely different results the second time, he was extremely puzzled. It was like doing the sum 4 + 4 twice and getting a different result each time. A closer examination revealed a vanishingly minute difference in the initial conditions of the simulation; the second time he had run the simulation, his input settings had been saved to three decimal places instead of six, so that the values had been very minutely rounded. So the differences in the initial conditions were of the order of the fourth decimal place. Yet the output of the simulation was dramatically different.

This scenario struck at the heart of classical Newtonian science, which had always been predicated on determinism, the philosophical belief that every event or action is the inevitable result of preceding events and actions. Even before Newton, the astronomer and mathematician Johannes Kepler had described the "machinery of the heavens" as being "like a clock," and since Newton had given his laws of motion and discovered the mathematics of gravity, the guiding principle of science had been that the universe is essentially mechanistic. The analogy most commonly used is of billiard balls colliding—if the motion, directions and force of the balls before collision is known, the same values for after the collision can be determined. This analogy applies particularly well to matter at the level of microscopic particles and atoms, which act like billiard balls zooming around, but can also be applied to, for instance, a planet going around the Sun. A core assumption of the deterministic

DETERMINISM UNDONE

model of the universe is that the more precisely you can measure the initial states of a system, the more precisely you can determine its outcome. The corollary of this is that in order to shrink uncertainty of outcome, all that is necessary is to measure more precisely the initial conditions. In his 1814 *Philosophical Essay on Probabilities*, the French mathematician Pierre-Simon Laplace asserted that if we knew everything about the universe in its current state, then "nothing would be uncertain and the future, as the past, would be present to [our] eyes."

What Lorenz had stumbled upon seemed clearly not to follow the logic of determinism. He had a found a system where, no matter how tiny the initial uncertainty, the outcome remained unpredictable. Grasping the significance of this discovery, he submitted his paper "Deterministic Nonperiodic Flow" to the *Journal of the Atmospheric Sciences* in 1963, pointing out that "slightly differing initial states can evolve into considerably different states." In fact, such systems had been known since at least 1890.

THE THREE BODY PROBLEM

As early as 1860 James Clerk Maxwell had modeled collisions between molecules and realised that tiny variations in initial states could be rapidly amplified to cause widely varying outcomes, and he had even proposed that free will, which seemed proscribed by a deterministic universe, might be explained through recourse to such a mechanism in the brain. In 1890 the French mathematician and physicist Henri Poincaré (1854–1912) discovered the phenomenon of dynamical instability. Newton's mechanics had been applied with amazing success to determination of planetary orbits around the Sun, examples of a two body problem (where one body orbits another). Tackling the Three Body Problem, relating to systems where three bodies are interacting gravitationally, Poincaré found that determinism seemed to break down. Whereas for two body problems, ever more precise measurement of the initial positions and movements of celestial bodies would lead to ever more precise predictions of their future positions, for three body problems the equations governing their motion meant that even the tiniest difference would propagate dramatically, so that it does not matter how small the initial uncertainty is, the uncertainty of outcome remains huge. Because it dealt with bodies in motion, this was known as dynamical instability; the modern term "chaos" dates to a 1975 paper on the phenomenon titled "Period Three Implies Chaos."

By this time chaos had already acquired its most popular label, the "butterfly effect," after the title of a talk given by Lorenz to the American Association for the Advancement of Science in 1972: "Predictability: Does the Flap of a Butterfly's Wings in Brazil Set off a Tornado in Texas?" In dynamically unstable systems like global weather, Lorenz suggested, an initial variable as small as the flap of a butterfly's wing could be magnified to produce a dramatic outcome like a tornado.

The butterfly effect is now recognized as a profoundly significant phenomenon. For a vast range of systems, only an infinitely precise specification of initial states, requiring an infinite number of measurements with infinitely precise instruments, can lead to a precisely determined outcome. In other words, they are indeterministic. This applies to the weather, neurological processes, animal populations, fluid dynamics, variability in the heart rate, oscillating magnetic pendulums, stock markets, and pseudo-random number generators, among others. In the case of the weather, chaos means that it will never be possible to give long-range forecasts with any accuracy.

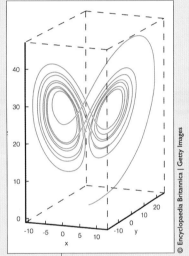

Physicists now believe that the universe may be essentially indeterministic, and it may be that chaos theory explains the arrow of time. The arrow of time is the phenomenon whereby time seems to run in one direction only, which is not a predicate of classical Newtonian mechanics. For instance, in a purely Newtonian universe, if billiard ball A collides with ball B so that it follows path C, then playing the shot in reverse so that B follows path C in reverse, should end up with ball A back exactly where it started. An observer looking at the situation at three different time points would have no way of knowing which state is the past and which the future. But chaos means that the shot can never be played perfectly in reverse, so that there can be no turning back the clock.

STRANGE ATTRACTOR
Lorenz also went on to discover an intriguing property of chaotic systems, known as the strange attractor, which demonstrates that chaotic systems need not be random or disordered, but can display regular patterns or cycles, such as that shown here.

SPINNING NEUTRON STARS: RADIO ASTRONOMERS DETECT PULSARS

1967

Discoverer: Jocelyn Bell Burnell and Tony Hewish

Circumstances: Poring over hundreds of feet of radio telescope readouts, searching for quasars, doctoral student Jocelyn Bell notices "a bit of scruff on the records"

Consequences: Discovery of a new class of astronomical object

I went home that evening very cross; here was I trying to get a Ph.D. out of a new technique, and some silly lot of little green men had to choose my aerial and my frequency to communicate with us.

Jocelyn Bell Burnell, "Little Green Men, White Dwarfs, or Pulsars?" (1977)

Perhaps the closest astronomers have ever come to thinking they had made contact with extraterrestrial civilization was in 1967, when a graduate student operating a crude radiotelescope array noticed that something—or someone—was broadcasting a repeating pulse of radio signals from far out in space.

In 1933 Karl Jansky, an engineer for Bell Telephone Laboratories, was assigned the task of identifying sources of static that might interfere with a proposed trans-Atlantic radio telephone service. Using a radio antennae array of his own design, Jansky found radio static generated by thunderstorms, but also a persistent low "hiss" that fluctuated on a day-night schedule, but did not come from the Sun. The signal repeated every 23 hours and 56 minutes, rather than exactly 24 hours, leading Jansky to realize that it followed sidereal time, which is to say it related to the fixed stars. Tracking the static to its source, he discovered that it emanated from the center of the Milky Way. Jansky had discovered cosmic radio waves. Although these may have been observed as early as the late 19th century by radio pioneers such as Tesla (see page 175), it was Jansky's discovery that inspired the development of radio astronomy, particularly after World War II.

In the mid-1960s, a new technique called interplanetary scintillation was developed that could pick out compact radio sources, such as quasars (now believed to be the central galactic regions around supermassive black holes). This prompted Cambridge University radio astronomer Professor Tony Hewish to design a large radio telescope array that would use IPS for a quasar survey of the sky. A young graduate student named Jocelyn Bell signed on to help, and in the summer of 1967 found herself working alongside volunteers to construct an antennae array covering an area of 4.5 acres, in which over a thousand posts were put up. Bell-Burnell, as she became after getting married, recalled, "We did the work ourselves—about five of us—with the help of several very keen vacation students who cheerfully sledge-hammered all one summer."

Once the array was operational in July, the signals it received were recorded by a pen-tracer on a continually unspooling pile of graph paper. It was Bell's job to analyze the records by hand. As the weeks

A BIT OF SCRUFF

RADIO STAR
Jocelyn Bell (Burnell, since her marriage), shown with a page from a graph readout, in front of a radio astronomy dish.

went by, Bell became adept at "reading" the hundreds of feet of chart, and around seven weeks after recording began, she noticed something unusual: "I became aware that on occasions there was a bit of 'scruff' on the records, which did not look exactly like a scintillating source, and yet did not look exactly like man-made interference either."

THE CRAB NEBULA
The remants of a supernova that exploded 1,000 years ago. At its heart lies a spinning neutron star, from which beams of radio waves sweep round 30 times a second, showing up on Earth's radio telescopes as a pulsar.

Bell realized that the same radio source had left its trace on the charts on previous occasions, and it was agreed that it was worth taking a closer look. This meant running the paper through the pen recorder more quickly, so that any peaks and troughs caused by fluctuating radio signals would be more widely spread out and visible in higher resolution. "At the end of November '67 I got it on the fast recording," Bell recalled. "As the chart flowed under the pen I could see that the signal was a series of pulses, and my suspicion that they were equally spaced was confirmed as soon as I got the chart off the recorder. They were 1.3 seconds apart." A regular series of sharp pulses at this rate immediately suggested an artificial source, but when Hewish came down to the array to see for himself it was apparent that the source followed sidereal time, and that it was not due to any obvious sources of interference, such as radar reflections off the moon. Other astronomers confirmed the pulses, and proved that they did not come from inside the solar system. But it seemed impossible that such rapid pulses could come from a star; a celestial object as huge as a star cannot "switch on and off" at such a rate, nor can it spin fast enough without flying apart from centrifugal forces. With no other explanation, the Cambridge astronomers were left contemplating the only other possibility; as Bell put it: "were these pulsations man-made but made by man from another civilization?"

LITTLE GREEN MEN

Bell and Hewish jokingly referred to their mystery radio source as Little Green Men or LGM, but they were left with a serious headache. "If one thinks one may have detected life elsewhere in the universe," Bell wondered, "how does one announce the results responsibly? Who does one tell first?"

In December Bell identified three other "lots of scruff," and in January 1968 the discovery of the first pulsing radio source was announced in a paper in the journal *Nature*. As soon as he heard about the discovery, the distinguished astronomer Fred Hoyle suggested that supernova remnants might be the cause, but when the press got hold of the report they focused on a line where Bell and Hewish mentioned "that at one stage we had thought the signals might be from another civilization." The added fact that a female astronomer had made the discovery sparked a press frenzy, and Bell was photographed in a variety of "eureka" poses—"Look happy dear, you've just made a Discovery!"— and quizzed on her love life and height compared to Princess Margaret.

Philip Morrison, Professor of Physics at the Massachusetts Institute of Technology, recalls meeting an astronomer friend returning from Britain in early 1968: "And he said, 'Have you heard the latest?' And I said, 'Well, what's that?' He said, 'They've got something that pulses every second—a stellar signal that pulses every second.' I said, 'Oh, that couldn't be true!' 'Yes,' he said, 'it's absolutely true... It's extraordinary.'" The paper in *Nature* sparked feverish activity in the world of astronomy. "Everybody with a radio telescope began looking in radio waves for pulsars," says Morrison. In 1968 another 27 pulsars were discovered (there are now over 1,000 pulsars known, out of an estimated 200,000 in our galaxy).

THE LIGHTHOUSE EFFECT

Perhaps the most significant pulsar discovery was made by English astronomer Mike Lodge working in Australia, who found a pulsar pulsing at the incredibly rapid rate of 11 pulses per second, discounting any possible mechanism whereby a normal star is somehow "switching" on and off very fast. The only explanation left was that the phenomenon must involve an incredibly tiny star—in other words the remnants of a supernova, just as Hoyle had suggested.

As early as 1933 Baade and Zwicky had suggested that the core of a massive supernova might collapse under such immense gravitational forces that even electrons and protons are crushed together to form neutrons, leaving essentially one giant nucleus—a colossal star crushed down into something just a mile or two across, which is spinning astronomically quickly. Lodge's discovery strongly suggested a potential candidate for the identity of pulsars, especially given that the pulsar he had found is located near the remnants of a supernova—exactly where a neutron star would be expected to be. So the discovery of pulsars had also helped to prove the reality of the formerly hypothetical neutron star.

"LUCY," THE HOMINID SKELETON: THE MISSING LINK IS FOUND
1974

Discoverer: Donald Johanson

Circumstances: American paleontologist excavating an Ethiopian site gets a "lucky feeling"

Consequences: The most complete hominid skeleton yet found completely rewrites the story of human evolution and becomes a global celebrity

Lucy was utterly mind-boggling.

Lucy: The Beginning of Humankind, Donald Johanson and Maitland Edey (1981)

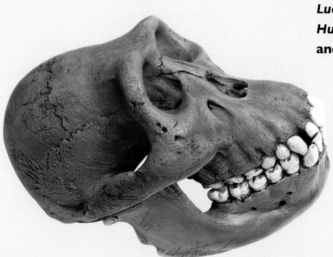

Lucy was the name given to the partial skeleton of an early human ancestor discovered at Hadar in the Afar region of Ethiopia in 1974. Still the most famous find in paleoanthropology (the study of prehistoric humans and human origins), Lucy dramatically overturned assumptions about human evolution.

In his brilliant and far-sighted 1871 book on human evolution, *The Descent of Man*, Charles Darwin pointed out that humanity's closest living relatives, the gorilla and the chimpanzee, are both confined to Africa, so that it would be logical to look there for an ancestor common to all three. In other words, Darwin controversially contended, humans evolved in Africa, probably millions of years ago, making every living person of ultimately African descent. Darwin's message was not well received by the primarily racist and colonialist prevailing discourse, with most scientists preferring to believe in a relatively recent Eurasian origin for humans dating back not much further than 100,000 years. Until the 1920s there was no fossil evidence from Africa for early hominins (humans and prehistoric species more closely related to humans than chimpanzees, a term not to be confused with hominids, a grouping that encompasses humans, other great apes, and their prehistoric ancestors), but there were fossils of Neanderthals from Europe and the human-like "Java man" from Southeast Asia (now classified as *Homo erectus*), found in 1891.

THE DESCENT OF MAN

In 1924 Raymond Dart discovered the Taung child in South Africa: a skull belonging to a young male with both ape-like and human anatomical features. Dart named the species of the Taung child as *Australopithecis africanus* ("southern ape from Africa"), and interpreted a crucial feature at the base of the skull to indicate that it walked upright. It seemed that Dart had found the missing link between humans and apes, but there was widespread resistance to this identification. Subsequent finds, however, cemented the status of the australopithecines as hominins, but important questions about human evolution remained. Were australopithecines on a side branch from the main line of descent from ape to humans? They had relatively small brains, not much bigger than a chimpanzee, and this did not fit with the prevailing view of human evolution, which assumed that the first step on the road to humanity was the evolution of a big brain. Only after this big brain enabled early hominins to invent tool use would they have started walking upright, to free up their hands, while the ability to exploit higher value sources of food in turn meant that they could come out of the forest and onto the savannah where their bipedalism provided even greater advantages. It was widely assumed that the breakthrough hominin fossil finds would be of this putative big-brained, pre-bipedal ape.

This view was strengthened by the discovery, in 1960, of the first humans: fossils of Homo *habilis* ("handy man," named for associated tools), dating to around 2 million years ago. This early transitional human fossil was discovered by Louis and Mary Leakey at Olduvai Gorge in Tanzania, part of the Great Rift Valley, a geological formation running up the eastern side of Africa, which is believed to have been the cradle for significant developments in human evolution. Similar fossils were discovered by the Leakeys' son, Richard, at Lake Turkana in Kenya in 1969. Specimens of *Homo habilis*, which was bipedal and made tools, show that early transitional humans had brains 35 percent larger, on average, than *Australopithecus africanus*. In other words, tool-making and bipedalism were indeed associated with larger brains in the first humans.

LUCKY LUCY

DONALD JOHANSON
Johanson has made or overseen many other major finds, including the first fossil hominid knee and a collection of bones from a family group known as the First Family.

To the north of the Leakeys' sites, in the semiarid Afar region of Ethiopia, American paleontologist Donald Johanson was leading a team digging at Hadar. Millions of years ago this area would have been wetter, with a lake and mixed grassland and forest. On November 30th, 1974, Johanson wrote in his daily diary that he had a good feeling. Filled with optimism he and graduate student Tom Gray set out in a Land Rover to spend a day surveying in the intense heat. On the way back to camp, a hunch led Johanson to stop off and look in the bottom of a nearby gully. "It had been thoroughly checked out at least twice before by other workers, who had found nothing interesting," he wrote in is 1981 bestseller *Lucy: The Beginning of Humankind*. "Nevertheless, conscious of the 'lucky' feeling that had been with me since I woke, I decided to make that small final detour."

Sure enough, Johanson spotted a bone: "I knew at once I was looking at a hominid elbow. I had to convince Tom, whose first reaction was that it was a monkey's." In fact it was the first inkling of probably the most amazing find in paleoanthropology to date. "Johanson had stumbled on a skeleton that was about 40 percent complete," wrote rival paleoanthropologists Richard Leakey and Roger Lewin, "something that is unheard of in human prehistory farther back than about a hundred thousand years. Johanson's hominid had died at least 3 million years ago."

In fact Johanson dated his fossil to 3.5 million years ago, making it both the most complete and the oldest hominid fossil then known. Officially designated as AL 288-1, the find consisted of skull fragments, a lower jaw, ribs, an arm bone, a portion of a pelvis, a thighbone, and fragments of shinbones. It appeared to belong to an adult female, and it soon acquired a famous nickname. Johanson explains: "The whole camp was listening to Beatles' tapes because I was a great Beatles fan, and 'Lucy in the Sky with Diamonds' was playing and this girl said, well if you think the fossil was a female, why don't you name her Lucy? Initially I was opposed to giving her a cute little name, but that name stuck." Johanson dubbed her species *Australopithecus afarensis* (after the Ethiopian region).

TWO LEGS GOOD

Along with her age and state of preservation, Lucy had another massive surprise in store. Her leg and hip bones clearly indicated that she walked upright on two legs, but her skull showed that her brain was little different in size to a chimpanzee. Lucy dramatically demonstrated that the received wisdom about the course of human evolution was back to front; bipedalism preceded brain size increase, and not vice versa. "She showed us conclusively," says Johanson, "that upright walking and bipedalism preceded all of the other changes we'd normally consider being human, such as tool-making." Against all expectations, it was bipedalism that set us on the path to becoming human.

Lucy had several adaptations indicating that she could and almost certainly did walk upright: compact feet to support her weight; a short, broad pelvis to hold her body upright; and thigh bones that angled inward to keep her body weight directly above her knees during a stride. She also had much more ape-like features. Along with her prominent jaw and small cranium, her funnel-shaped ribcage, and broad pelvis suggest that she had a large belly, like a modern ape,

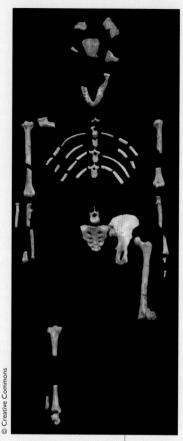

LUCY'S BONES
The partial skeleton of the young female *Australopithecus afarensis* that Johanson dubbed "Lucy." It may not look like much, but this degree of completeness is unprecedented in such an old paleoanthropological find.

probably reflecting the large, long intestines needed to digest a low-quality, bulky, mainly vegetarian diet. With a thick waist and high shoulders, she probably could not swing her arms like modern humans do when running, while her long, curved toes and strongly muscled arms and chest suggest that she retained adaptations for tree climbing. In other words, Lucy was very much a transitional stage in human evolution, adapted to exploit a mixture of habitats, including forest canopies and lakeside grasses.

Lucy rewrote the textbooks and made Johanson a star. "I had written papers about her, appeared on television, made speeches. I had shown her proudly to a stream of scientists from all over the world. She had—I knew it—hauled me up from total obscurity into the scientific limelight." The combination of her nickname, academic significance, and completeness made Lucy uniquely accessible and appealing as an emblem of human evolution and human origins. It is easy to picture her: a small, slight ape-woman, walking upright as she forages for food on the shores of a lake into which she will fall and perish. Lucy has exerted a profound influence on the wider public and academics alike, particularly on Johanson himself. "Lucy had been mine for five years," he recounted in his 1981 book, "The most beautiful, the most nearly complete, the most extraordinary hominid fossil in the world."

LIFE IN THE DEPTHS: LIFE-FORMS ARE FOUND AROUND HYDROTHERMAL VENTS

1977

CATEGORY

Astronomy

Biology

Chemistry

Exploration

Mathematics

Medicine

Physics

Technology

Discoverer: Galápagos Hydrothermal Expedition

Circumstances: A team of scientists equipped with submersible technology explores the seabed near the Galápagos Rift

Consequences: Discovery of a new geological phenomenon and new ecosystems

We all started jumping up and down. We were dancing off the walls. It was chaos. It was so completely new and unexpected… It was a discovery cruise. It was like Columbus.

John Edmond, MIT geochemist, quoted in the book *Water Baby* by Victoria Kaharl (1990)

© Hlewk | Dreamstime.com

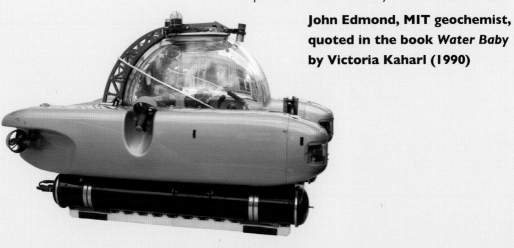

Geologists had suspected that hydrothermal vents must exist, but they were completely unprepared for the astonishing biological discoveries that accompanied them—a whole new type of life, with profound implications for the origins of life on Earth and the possibility of life on other planets.

HOT BRINES OF THE RED SEA

In the 1960s geologists pieced together the theory of plate tectonics, arriving at an understanding of the dynamic forces that shape the Earth's crust. The crust is made up of tectonic plates, and these move about because they are being driven apart at "spreading centers," where molten rock or magma from the mantle rises up and spills out, generating new ocean floor spreading out from ridges running the length of crust boundaries. As this new ocean floor cools from molten rock into cold basalt, it contracts and cracks, and scientists reasoned that seawater seeping down into these cracks would be heated by the magma below, rising back up as subsea hot springs, aka hydrothermal vents. Similar processes result in the geysers and hot springs of places like Yellowstone National Park. But the ocean floor remained a remarkably alien and unknown realm—even today more is known about the surface of the Moon than the bottom of the oceans. Did these hydrothermal vents really exist, and if so, where were they?

Scientists already possessed tantalizing clues thanks to a strange phenomenon observed in the Red Sea, which constitutes the northern end of Africa's Great Rift Valley. In the 1880s a Russian ship anchored off Mecca had taken a water sample from 650 yards (600 meters) down and discovered that it was warmer than the surface and extremely salty. This bizarre finding was replicated by a Swedish ship 40 years later. Initially it was believed that evaporation by the tropical sunlight had resulted in hot brines: strongly saline water, which, being denser, sank to the bottom of the Red Sea. In 1964 and 1965 samples taken by research vessels found seawater of over 122°F (50°C). In 1966 the US National Science Foundation sent a research vessel from the Woods Hole Oceanographic Institution (WHOI), which took a sediment core from the sea bottom. Scientists on board were astonished by what they saw. "The color variation is fantastic: all shades of white, black, red, green, blue, or yellow can be observed," wrote Egon Degens and David Ross. "Perhaps some of the more colorful Indian sand paintings and Mexican rugs faintly match these sediments in the variation and intensity of their colors."

The Red Sea hot brines and other evidence convinced the scientific community that they needed to launch a concerted effort to locate hydrothermal vents. In 1974 the French and the Americans combined forces in Project Famous, utilizing some of the most cutting-edge deep sea submersibles available, including Woods Hole's *Alvin*, a three-man piloted submersible, and *ANGUS*, a deep-towed camera sled. Despite successfully surveying the Mid-Atlantic Ridge, Project Famous failed to find a hydrothermal vent, although in one dramatic incident in 1975 *Alvin* was piloted into a huge fissure in the ocean floor, only for the pilots to realize they were stuck. Only by retracing their steps exactly did they get free and avoid certain death.

In 1972 and 1976, expeditions to the Galápagos Rift, a spreading plate boundary zone in the Pacific near the Galápagos Islands, had revealed hints of hydrothermal activity, including photographs of a patch of dead clams. In February 1977, the Galápagos Hydrothermal Expedition (GHE), funded by the National Science Foundation, sailed into the area to launch the most sophisticated search yet. The first step was to survey the target area of ocean floor with the deep-tow sled *ANGUS*, which took a photo every 10 seconds as it was dragged back and forth, and could detect even tiny changes in water temperature. On the evening of February 15th, around midnight, *ANGUS* registered a three-minute period of elevated water temperature. When it came back on board and its photographic payload was developed, scientists paid particular attention to the images corresponding to the warm water "temperature anomaly."

"The photograph taken just seconds before the temperature anomaly showed only barren, fresh-looking lava terrain," recalled oceanographer Bob Ballard in a 1977 article in *Oceanus* magazine, "Notes on a Major Oceanographic Find." "But for 13 frames (the length of the anomaly) the lava flow was covered with hundreds of white clams and brown mussel shells. This dense accumulation, never seen before in the deep sea, quickly appeared through a cloud of misty blue water and then disappeared from view. For the remaining 1,500 pictures, the bottom was once again barren of life."

OASIS OF LIFE

WHITE SMOKER
The Champagne Vent, a hydrothermal vent on the flanks of an undersea volcano in the Pacific Ocean, which jets liquid carbon dioxide into the water. The vent fluid does not contain metals, which oxidize on contact with seawater to produce the black plumes characteristic of "black smoker" vents.

Within hours scientists were on board *Alvin* and heading for the anomaly region. "When they reached their target coordinates," wrote Ballard, "*Alvin* and its three-man crew entered another world." They saw water shimmering with heat coming up through cracks in the cooled lava of the sea bed, which immediately turned a cloudy blue as the water cooled and dissolved minerals such as manganese precipitated out and drifted down to stain the lava a rusty brown. *Alvin*'s temperature sensors showed the water was 46°F (8°C) —they had found the long sought for hydrothermal vent. More amazing still, the scientists saw "a dense biological community living in and around the active vents," made up of large animals, including big white clams 1 foot (30 centimeters) across. On board *Alvin*, Jack Corliss of Oregon State University exclaimed, "Isn't the deep ocean supposed to be like a desert? Well, there's all these animals down here." The GHE scientists had discovered something unprecedented and entirely unexpected. "This oasis of life was only 50 meters [165 feet] across and totally different from that of the surrounding area," Ballard reflected, "What were the organisms eating? They were living on solid rock in total darkness."

When *Alvin* returned to the surface and the scientists opened up the water samples taken from the vent region, the smell of rotting eggs drove them to the portholes, but also solved the mystery. The water was rich in chemically active hydrogen sulfide, which provided energy to bacteria that formed the bottom of a food chain entirely independent of sunlight. Further dives revealed two other vent sites teeming with life, including one with large white-stalked tubeworms with bright red tops and strange organisms resembling dandelions, which they dubbed "the Garden of Eden."

Onboard the research vessels there was great excitement. MIT geochemist John Edmond recalled that "We were dancing off the walls. It was chaos. It was so completely new and unexpected that everyone was fighting to dive... It was like Columbus." On board the ship was a journalist for the *San Francisco Chronicle*. "Astounding Undersea Discoveries," read his headline. "When these findings are all analyzed in detail they are bound to 'revolutionize' many theories about the deep ocean floor."

The GHE team had included no biologists, but now there was feverish excitement in the biology community and in 1979 a shipload of them arrived at the Galápagos Rift. "We would subject the newly discovered communities to the full arsenal of techniques available to modern biology," wrote expedition leader J. Frederick Grassle, then a scientist at Woods Hole Oceanographic Institution. They found an incredible bounty of new life-forms unknown to science, including mussels, anemones, whelks, limpets, feather-duster worms, snails, lobsters, brittle stars, and blind white crabs. "Literally every organism that came up was something that was unknown to science up until that time," said Richard Lutz, then a post-doctoral scientist at Yale, now a professor at Rutgers University. "It made it terribly exciting. Anything that came [up]… was a new discovery."

The most impressive organisms, red-tipped tubeworms that grew 8 feet (2.5 meters) tall, proved to have no mouths and no guts. Instead they survive through symbiosis with hydrogen sulfide processing bacteria. Many other hydrothermal vent organisms have similar arrangements. Writing in the *Annual Review of Microbiology*, Woods Hole biologist Holger Jannasch concluded: "We were struck by the thought, and its fundamental implications, that here solar energy, which is so prevalent in running life on our planet, appears to be largely replaced by terrestrial energy—chemolithoautotrophic bacteria taking over the role of green plants. This was a powerful new concept and, in my mind, one of the major biological discoveries of the 20th century."

Subsequent expeditions discovered black smoker vents where super-heated water at over 662°F (350°C) blasts out of the seafloor, creating huge chimneys of mineral deposits. Even here life can flourish, and scientists now believe that these extreme ecosystems, which rely entirely on geothermal energy rather than sunlight, could be where life on Earth first evolved. They also show us what life on other planets might well look like, and it is believed that hydrothermal vents on Europa, the icy moon of Jupiter, might be the best bet for detecting extraterrestrial life in our solar system.

BLACK SMOKERS

© AF archive | Alamy

TUBE WORMS
These bizarre plant-like forms are actually worms. They build tubes of calcium carbonate for protection, with fern-like "plume" organs projecting out of them. These plumes are bright red because they are rich in blood, circulating through them to extract nutrients from the sea water.

CATEGORY

Astronomy

Biology

Chemistry

Exploration

Mathematics

Medicine

Physics

Technology

THE FIRST EXOPLANET: PLANETS ARE OBSERVED CIRCLING OTHER SUNS

1995

Discoverer: Michel Mayor and Didier Queloz

Circumstances: Astronomers at a French observatory switch on a new instrument, designed to look for minute stellar wobbles, and find one almost immediately

Consequences: Opens the floodgates to a swelling number of exoplanet discoveries

The search for extrasolar planets can be amazingly rich in surprises.

Michel Mayor and Didier Queloz, "A Jupiter-mass companion to a solar-type star," *Nature*, November 1995

In 1584 the Catholic monk Giordano Bruno suggested that there are "countless suns and countless earths all rotating around their suns." Bruno was burned at the stake for his heresy, but he was not the first or last to suggest as much. Until very recently, however, the possibility of detecting extrasolar or exoplanets—planets orbiting suns other than our own—seemed impossibly remote. Yet in the space of two decades scientists have now detected hundreds, completely changing our view of the universe and the possibility of extraterrestrial life.

Major schools of philosophy in ancient Greece asserted that there must be other worlds orbiting other suns. From the 5th century BCE the atomists and their successors the Epicureans argued for a plurality of worlds in the cosmos. In around 430 BCE Leucippus described how "worlds come into being," imagining "many bodies of all sorts and shapes" moving "colliding with one another" and separating "like to like." Around 380 BCE Democritus described strange planets very different to Earth: "In some worlds there is no Sun and Moon, in others they are larger than in our world, and in others more numerous... There are some worlds devoid of living creatures or plants or any moisture." About a century later Epicurus claimed, "There are infinite worlds both like and unlike this world of ours. For the atoms being infinite in number, as was already proven [...] there nowhere exists an obstacle to the infinite number of worlds." Yet in the post-Classical period these voices were drowned out by the authority of Aristotle. "There cannot be more worlds than one," he argued, asserting the Earth to be at the center of revolving crystalline spheres carrying the other bodies in the solar system, surrounded by an unchanging sphere of fixed stars. In other words, there were no other suns and no room for other planets.

Aristotle's philosophy became the prevailing dogma of medieval Christianity and scholarship (see page 53), and so the very notion of extrasolar planets became heretical, as Bruno learned to his cost. His terrible fate helped convince Galileo to abjure his own claims about the nature of the planets and the nongeocentric nature of the cosmos, but the reality revealed by his telescope (see page 58) could not be hidden once unveiled. Natural philosophers began to realize that the universe was much bigger than previously believed, with much more in it. "Why may not every one of these Stars or Suns have as great a Retinue as

MORE WORLDS THAN ONE

our Sun, of planets, with their moons, to wait upon them?" pondered Christiaan Huygens in his 1698 book, *Cosmotheros*, which explored the concept of life on other planets. Discoveries of trans-Saturnian planets showed that as the power of telescopes increased previously unimaginable bodies might become visible (see page 102). When it became clear that there are not only millions of stars in our galaxy, but millions of galaxies in the universe, astronomers realized that there almost certainly must be extrasolar planets out there; it was simply a matter of finding them.

A MOTH NEAR A SPOTLIGHT

Terrestrial astronomers have to surmount immense obstacles to even have a chance of detecting exoplanets. Planets do not generate their own light (except in the low energy infrared end of the spectrum), they only reflect the light of their sun(s), but they are astronomically dimmer. In our own solar system, for instance, the Sun is a billion times brighter than Jupiter and 10 billion times brighter than the Earth. This makes it incredibly hard to spot a dim planet against the glare of its sun. For example, a planet orbiting the nearest star to our own, Proxima Centauri, would be 7,000 times more distant than Pluto, so trying to observe it directly would be like standing in Boston and looking for a moth near a spotlight in San Diego.

© Franck Schneider | Creative Commons

MICHEL MAYOR
Swiss astrophysicist Mayor's work at the Geneva Observatory pioneered important techniques in extrasolar planet hunting; he went on to lead a team searching for smaller, more Earth-like exoplanets at the European Southern Observatory telescope in Chile.

Astronomers can observe stars, however, and due to the gravitational interaction between them, planets have an effect on stars. Although we commonly speak of a planet orbiting a sun, in truth any two bodies orbit a common center of gravity. Because stars are so much bigger than their satellites, this common center of gravity is likely to be within the radius of the star, but since it is not exactly at the center, the star will wobble as the planet orbits around it. Astronomers' ability to detect such effects depends on the precision and resolution of their instruments. In the 1960s the Dutch astronomer Peter van de Kamp claimed that he could detect, using a technique called astrometry, wobbles in the orbit of Barnard's star that indicated that it had a planetary system. Astrometry (literally "star measurement") involves comparing photographs of a star taken at different times to see if it has moved (as when wobbling), but the precision needed to detect such minute changes was beyond that available to van de Kamp and his claims have been discounted.

The periodic radio pulses from a pulsar offer the necessary precision, however, and the first detection of extrasolar planets was achieved by measuring minute deviations in the regularity of pulses coming from PSR B1257+12, a pulsar in the constellation Virgo. In 1992 Aleksander Wolszczan and Dale Frail discovered tiny anomalies in the timing of its pulses, showing that these came from the existence of two small rocky planets orbiting the pulsar. Because pulsars are created after a supernova explodes and collapses, the planets must have been formed or captured after the star became a pulsar, and because they are constantly blasted with vast quantities of high energy radiation from the pulsar, they must be utterly lifeless. The peculiarity, specificity, and inhospitability of this pulsar exoplanet system means that Wolszczan and Frail are, perhaps unfairly, not usually credited with the first discovery of an extrasolar world.

Another way to detect the stellar wobbles caused by orbiting planets is to measure the radial velocity of a star, which is its motion toward and away from the Earth as it wobbles. Such motion affects the wavelength of the light that reaches us through the Doppler effect, the phenomenon whereby a moving source causes compression or stretching out of waves, such as the change in pitch heard in a siren as an ambulance moves toward and then past a listener. Doppler shifts caused by stellar wobbles can be measured using spectrometers, but the wobbles are extremely tiny. In our solar system, for instance, Jupiter causes the Sun to change velocity by about 13 yards/second (12m/s). In the early 1980s the best resolution achievable was around 0.6 miles/second (1km/s). But technological advances brought this figure down to around 22 yards/second (20m/s) by the mid-1990s. This made it possible to discover by this technique comparatively large wobbles, such as are induced by a large planet bigger than Jupiter. It was such a planet that Michel Mayor and Didier Queloz were hoping to snare in 1994 when they turned on their new instrument, the ELODIE spectrograph, at the Observatoire de Haute-Provence in France.

HOT JUPITER

© M.McCaughrean (ESA)/ESO | Creative Commons

DIDIER QUELOZ
Swiss astronomer Queloz went on to become one of the world's leading experts on extrasolar planets and a prolific planet hunter.

Planet hunting was, Queloz recalled on being awarded the BBVA Foundation Frontiers of Knowledge Award in 2011, "a kind of bizarre, weird project." He and Mayor were not expecting instant results: "We had built this really precise machine and thought it was going to take years to detect a planet, then suddenly after a couple of months, there was the first signal." Observing the star 51 Pegasi, they saw a radial velocity of around 50 meters/s; indicative of a Jupiter-size planet, presumably a gas giant. But there was a problem: "What we were observing didn't fit with any known planet in the solar system."

By looking at how often the wobble occurs, it is possible to measure the period of the orbit and thus work out how close to its star a planet is orbiting. What they found was that the massive gas giant is going round its sun once every four days, which means it is six times closer to 51 Pegasi than Mercury is to the Sun. "At first I thought I was mistaken," recalls Queloz, "but Michel has the kind of mind that is ready for the unexpected, and that was essential to our success." They waited a year until they could observe the star again, and "in July '95 we repeated our measurements and got exactly the same signal. It was then we knew that we had really a planet." The gas giant they found—called 51 Pegasi b—must have surface temperatures of 1,832°F (1,000°C), leading to it being dubbed a "hot Jupiter."

The two astronomers submitted a paper for a November issue of *Nature*, but announced their findings at a congress in Florence in October. They were unprepared for the global excitement that resulted. "It was a completely crazy time, with calls from papers, from television, from radio, from all the world," they later recalled. "It was only then we realized how important our work was for the general public," said Mayor.

In the wake of Mayor and Queloz's discovery hundreds more exoplanets have been discovered including some near-Earth-sized rocky planets orbiting within the habitable zones of their solar systems (the optimal orbital range for getting enough but not too much radiation). Astronomers have been able to directly image extrasolar planets and with the help of future space telescopes it may be possible to look for traces of life in the atmospheres of worlds orbiting other suns. Thanks to the planet hunters, we may yet discover extraterrestrial life.

THE HIGGS BOSON: THE GOD PARTICLE

2012

CATEGORY

Astronomy

Biology

Chemistry

Exploration

Mathematics

Medicine

Physics

Technology

Discoverer: Large Hadron Collider

Circumstances: A vast particle accelerator is built to test a prediction made 50 years earlier; rival teams of scientists race to find the telltale traces of an elusive particle

Consequences: Confirms the Standard Model of physics

This boson is so central to the state of physics today, so crucial to our final understanding of the structure of matter, yet so elusive...

Leon Lederman, *The God Particle: If the Universe Is the Answer, What Is the Question?* (1993)

This book has tried to focus on "eureka" moments, when a previously unknown or even unguessed at truth is dramatically revealed at a stroke to a lone explorer. In some senses the "discovery" of the Higgs boson is the antithesis of this: a long predicted theoretical concept is proven to most likely exist thanks to the cumulative evidence of billions of experiments carried out by teams of thousands of scientists working on a vast machine specifically built for the purpose. Yet it is widely recognized as the discovery of the century so far, and an exemplar of discovery in the age of "big science."

THE GOD PARTICLE

The Higgs boson is a particle that mediates the effects of the Higgs field, a concept characterized in the mid-1960s by several physicists, most famously by Peter Higgs of the University of Edinburgh. In 1964 he wrote a seminal paper in the journal *Physical Review Letters*, "Broken Symmetries and the Masses of Gauge Bosons." The Higgs field is an attempt to explain why/how things have mass. It is a pervasive field

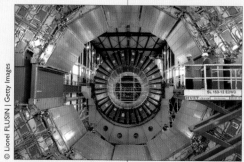

that interacts with different fundamental particles in different ways to give them mass. Several analogies have been used to explain it. The best known is that the Higgs field is like a crowd of people at a party, spread out fairly evenly until a celebrity arrives, whereupon they crowd around the star, giving momentum and inertia. Another analogy is that the Higgs field is like a snow field, across which move analogies of different particles, slowed to a greater or lesser degree by the snow: electrons, which have very tiny mass, are like skiers; fairly massive protons and neutrons are like people on snowshoes; very massive W and Z bosons are like men in heavy boots; massless photons are like birds who don't have any contact with the snow.

MIGHTY MAGNET
A vast magnet, the largest superconducting solenoid ever constructed, is slotted into place during the construction of the colossal CMS detector at CERN's large hadron collider.

The Higgs field was prefigured by the ancient Greek Stoic philosophers with their concept of *pneuma*, a state of tension that permeates the universe providing cohesion and substance to bodies. According to Joe Incandela, the lead scientist of the Compact Muon Solenoid (CMS) experiment at the large hadron collider (LHC), "The Higgs is sort of like the mother of everything… It tells you something very fundamental about the entire universe. So measuring its mass, for instance, could tell us whether the universe is stable or not... [it] is everywhere. Throughout the universe. So for me that is a really profound thing about the Higgs. It's not like other particles."

The Higgs boson should bear the names of several other scientists. Tom Kibble of Imperial College, London, Carl R. Hagen of the University of Rochester and Gerald Guralnik of Brown University all contributed to the theory, while Belgian physicists Robert Brout and François Englert at the Free University in Brussels actually published a paper on the same field *before* Higgs. "In the spring of 1964 we were both extremely excited," recalled Brout. "For the first time in my life, I felt what it might be to be a great physicist."

However, "the Brout-Englert-Higgs boson" is a mouthful, and proved a lot less popular than the nickname coined in 1993 by Nobel Prize-winning physicist Leon Lederman in his book *The God Particle*. "This boson is so central to the state of physics today, so crucial to our final understanding of the structure of matter, yet so elusive..." he wrote in his book, continuing: "Why God Particle? Two reasons. One, the publisher wouldn't let us call it the Goddamn Particle, given its villainous nature and the expense it is causing. And two, there is a connection, of sorts, to another book, a much older one..." Higgs has admitted he is uncomfortable with the nickname. "I wish he hadn't done it," he told the *Guardian* newspaper. "I have to explain to people it was a joke."

The theories of Higgs and co. caused excitement in physics because they offered to fill a major void in the Standard Model, the set of mathematical and theoretical constructs built up to explain the nature of the universe, including electromagnetism and radioactivity, and particles such as electrons, protons, photons, and their constituents. Missing from the Model was a way to include gravity. Proving that the Higgs boson really exists would confirm the validity of the Standard Model, and by extension validate most of the physics research of the last hundred years. What constitutes proof? Using the equations of Higgs and the others, physicists could predict the likely energy level of the theoretical particle. If they could smash together fundamental particles hard enough to create new particles, they could look at the energy signature of the new particles to see if they matched the predictions. But the energy levels involved are colossal, mirroring those

THE BIGGEST MACHINE IN THE WORLD

experienced less than a trillionth of a second after the Big Bang. To achieve this, it is necessary to build huge colliders that can accelerate protons to near light speeds. Teams in Europe and the United States raced to do just this.

The European Organization for Nuclear Research in Geneva, known as CERN, built the Large Electron-Positron Collider (LEP), which operated from 1989 to 2000. In the summer of 2000, just as it was about to be shut down for replacement, LEP caught glimpses of what some scientists believed to be the Higgs, but when the switch off was postponed until November, the glimpses proved illusory. Meanwhile at the US Fermilab a collider called the Tevatron was also Higgs hunting, but despite unfounded rumors of discovery in 2006, the Tevatron was not quite powerful enough. The field was left clear for LEP's replacement, the colossal LHC, which can accelerate protons to 99 percent of the speed of light. Protons race 11,245 times a second around a 17-mile (27-kilometer) long loop underneath the Swiss and French countryside, before smashing into each other to create tiny fireballs that recreate the conditions of the universe a trillionth of a second after the Big Bang. According to the theory, a Higgs boson is formed once in every four billion such collisions. Costing around $10 billion, the LHC was turned on in 2008. "It'll be a relief for them to find it..." commented Higgs shortly before the machine began searching for his boson. "If I'm wrong, I'll be rather sad." "I knew that when the Large Hadron Collider turned on, it was going to change us in some very dramatic way," commented David Kaplan, theoretical particle physicist at Johns Hopkins University and one of the men behind the 2014 documentary *Particle Fever*.

Two huge detectors—the ATLAS and the CMS—independently searched for the Higgs boson, with teams of thousands of scientists on each experiment. Both detectors observe billions of collisions and measure the energy signatures of particles produced, looking for statistically significant patterns. When a potentially meaningful anomaly arises, it usually proves to be a meaningless fluctuation as more data is accumulated. "We've made many discoveries, most of them false," remarked Guido Tonelli, Incandela's predecessor on the CMS, "But you never know when one of them will change everything."

After several false starts, an anomaly initially detected in May 2011 came to look more and more significant. On November 8th Tonelli was briefed on the improving indications. "It was my birthday. I considered it a sort of gift," he recalled. The news was relayed at a CERN briefing in December, and it was agreed that the definitive judgment would be handed down to coincide with the next major physics meeting, a conference in July 2012. To avoid unintentionally biasing the data gathering, the LHC teams switched to a "blind analysis" mode, where they would not have access to the results until mid-June. When the embargo was lifted, 6,000 ATLAS and CMS physicists devoured them in a frenzy of analysis. A major press conference was scheduled for July 4th, with Peter Higgs and other senior figures invited to attend. "We have observed a new boson," announced Joe Incandela, "This is very preliminary result, but it's very strong." The official declaration fell to the Director General of CERN, Rolf-Dieter Heuer, who told the world "I think we have it."

In 2013 the Nobel Prize for Physics was awarded to Peter Higgs and François Englert, "for the theoretical discovery of a mechanism that contributes to our understanding of the origin of mass of subatomic particles, and which recently was confirmed through the discovery of the predicted fundamental particle, by the ATLAS and CMS experiments at CERN's Large Hadron Collider." "There are 6,000 Higgs soldiers," said Eilam Gross, one of the leads on the ATLAS team, referring to the army of LHC physicists, "and they all deserve the Nobel Prize."

WE HAVE A BOSON

FURTHER READING

Flint-knapping

Coolidge, Frederick, and Thomas Wynn. *The Rise of Homo Sapiens—The Evolution of Modern Thinking.* Malden, MA; Oxford, UK: Wiley-Blackwell, 2009.

Rudgley, Richard. *Lost Civilizations of the Stone Age.* London, UK: Arrow, 1998.

Wayman, Erin. "Becoming Human: The Origin of Stone Tools." www.smithsonianmag.com

Taming the Flames

Adler, Jerry. "Why Fire Makes Us Human." *Smithsonian Magazine,* June 2013.

Roebroeks, Wil, and Paulo Villa. "On the Earliest Evidence for Habitual use of Fire in Europe." Proceedings of the National Academy of Sciences Early Edition, March 2011.

Wrangham, Richard. *Catching Fire: How Cooking Made Us Human.* New York, NY: Basic Books, 2009.

The Pythagorean Theorem

Diggins, Julia. *String, Straight-edge, and Shadow.* New York, NY: Viking Press, 1965.

Levy, Joel. *A Curious History of Mathematics.* London, UK: Andre Deutsch, 2013.

Newman, James Roy. *The World of Mathematics, Vol 1.* London, UK: George Allen & Unwin Ltd, 1960.

The Story of Mathematics: storyofmathematics.com

The Archimedes Principle

Bragg, Melvyn. *On Giants' Shoulders: Great Scientists and Their Discoveries from Archimedes to DNA.* London, UK: Sceptre, 1999.

Hirshfield, Alan. *Eureka Man: The Life and Legacy of Archimedes.* New York, NY: Walker & Company, 2013.

An Explosive Reaction

Kelly, Jack. *Gunpowder: Alchemy, Bombards & Pyrotechnics: The History of the Explosive that Changed the World.* New York, NY: Basic Books, 2004.

Needham, Joseph. *Science and Civilization in China: Military Technology, The Gunpowder Epic.* Cambridge, UK: Cambridge University Press, 1974.

Partington, J. R. *A History of Greek Fire and Gunpowder.* Baltimore, MD: Johns Hopkins University Press, 1960.

The New World

Bergreen, Laurence. *Columbus: The Four Voyages, 1492–1504.* Harmondsworth, UK: Penguin, 2013.

Fernandez-Armesto, Felipe. *1492: The Year our World Began.* London, UK: Bloomsbury, 2010.

Hamilton, Neil. *Scientific Exploration and Expeditions: From the Age of Discovery to the Twenty-First Century.* New York, NY: Sharpe Reference, 2010.

By Sea to the Indies

Clark Northrup, Cynthia, ed. *Encyclopedia of World Trade: From Ancient Times to the Present.* New York, NY: Sharpe, 2005.

Cliff, Nigel. *The Last Crusade: The Epic Voyages of Vasco da Gama.* London, UK: Atlantic, 2012.

Dear, I. C. B., and Peter Kemp, eds. *The Oxford Companion to Ships and the Sea (2 ed.).* Oxford, UK: Oxford University Press, 2006.

Into the Jungle

Levy, Buddy. *River of Darkness: Francisco Orellana's Legendary Voyage of Death and Discovery Down the Amazon.* New York, NY: Bantam, 2011.

Levy, Joel. *Lost Histories.* London, UK: Vision, 2007.

A Star is Born

Boerst, William J. *Tycho Brahe: Mapping the Heavens.* Greensboro, NC: Morgan Reynolds Publishing, 2003.

Gow, Mary. *Tycho Brahe Astronomer.* Berkeley Heights, NJ: Enslow Publishing, 2002.

Thoren, Victor E. *The Lord of Uraniborg: A Biography of Tycho Brahe.* New York, NY: Cambridge University Press, 1990.

Challenging Aristotle

Devreese, Josef, T. and Berghe Vanden. *Guido Magic is No Magic: The Wonderful World of Simon Stevin.* Southampton, UK: WIT Press, 2008.

Gravity Probe B: Testing Einstein's Universe: http://einstein.stanford.edu

The Galileo Project: http://galileo.rice.edu

Courting Controversy

Donovan, Charles J. ,trans. *Galileo's Observations of Jupiter's Moons.* www.dioi.org/galileo/galileo.htm

Heilbron, J. L. *Galileo.* Oxford, UK: Oxford University Press, 2010.

Wooton, David. *Galileo: Watcher of the Skies.* New Haven, CT; London, UK: Yale University Press, 2010.

Super Fast Math

Clark, Kathleen M., and Clemency Montelle. "Logarithms: The Early History of a Familiar Function." *Loci,* January 2011.

Gladstone-Millar, Lynne. *John Napier: Logarithm John.* Edinburgh, UK: National Museums of Scotland Publishing, 2003.

Domestic Annals of Scotland;:Reign of James VI. 1591–1603 Part C: www.electricscotland.com/history/domestic/vol1ch8c.htm

Storming the Darkness of the Mind

Caspar, Max, trans. and Hellman, C. Doris, ed. *Kepler.* London, UK; New York, NY: Abelard-Schuman, 1959.

Sobel, Dava. *A More Perfect Heaven.* London, UK: Bloomsbury, 2011.

Abhorrent to Nature

Meli, Domenico Bertoloni. *Thinking with Objects: The Transformation of Mechanics in the Seventeenth Century.* Baltimore, MD: Johns Hopkins University Press, 2006.

Middleton, William Edgar Knowles. *The History of the Barometer.* Baltimore, MD: Johns Hopkins University Press, 1964.

An Odd Straying of the Light

Hall, Rupert. A. "Beyond the Fringe: Diffraction as Seen by Grimaldi, Fabri, Hooke, and Newton." *Notes and Records of the Royal Society of London,* Vol. 44, No. 1, January 1990.

Francesco M. Grimaldi and His Diffraction of Light: www.faculty.fairfield.edu/jmac/sj/scientists/grimaldi.htm

Newton's Genius

Gjertsen, Derek. *The Newton Handbook,* London, UK; New York, NY: Routledge & Kegan Paul, 1986.

Gleick, James. *Isaac Newton.* London, UK: Fourth Estate, 2003.

Levy, Joel. *Newton's Notebook.* Stroud, UK: The History Press, 2009.

A Wonderful Spectacle

Ford, Brian J. "From Dilettante to Diligent Experimenter: A Reappraisal of Leeuwenhoek as Microscopist and Investigator." *Biology History,* 5 (3), December 1992.

Huxley, Rob. *The Great Naturalists.* London, UK: Thames & Hudson, 2007.

Ruestow, Edward, G. *The Microscope in the Dutch Republic: The Shaping of Discovery.* Cambridge, UK: Cambridge University Press, 1996.

The Great Southern Land

Withey, Lynne. *Voyages of Discovery: Captain Cook and the Exploration of the Pacific.* Berkeley and Los Angeles, CA: University of California Press, 1989.

The Captain Cook Society: www.captaincooksociety.com

Captain Cook's Journal: http://captainjamescook.wordpress.com

Oxygen

Johnson, Stephen. *The Invention of Air: An Experiment, a Journey, a New Country and the Amazing Force of Scientific Discovery.* London, UK: Penguin, 2009.

Levy, Joel. *The Bedside Book of Chemistry.* Sydney, Australia: Murdoch Books Pty Limited, 2011.

Priestley, Joseph. "Collected Essays," Vol. III. *Macmillan's Magazine,* 1874.

The Georgian Planet

Holmes, Richard. *The Age of Wonder.* London, UK: HarperCollins, 2008.

Hoskin, Michael. *Discoverers of the Universe: William and Caroline Herschel.* Woodstock, NY: Princeton University Press, 2011.

Lemonick, Michael D. *The Georgian Star: How William and Caroline Herschel Revolutionized Our Understanding of the Cosmos.* New York, NY; London, UK: W. W. Norton, 2009.

Animal Electricity

Baigrie, Brian. *Electricity and Magnetism: A Historical Perspective.* Westport, MA: Greenwood Press, 2007.

Fisher, Len. *Weighing the Soul: Scientific Discovery from the Brilliant to the Bizarre.* New York, NY: Arcade Publishing, 2011.

Malmivuo, Jaakko, and Robert Plonsey. *Bioelectromagnetism: Principles and Applications of Bioelectric and Biomagnetic Fields.* Oxford, UK: Oxford University Press, 1995.

Vaccination

Riedel, Stefan. "Edward Jenner and the History of Smallpox and Vaccination." *Baylor University Medical Centre Proceedings,* 18 (1): 21–25, Jan 2005

Dr. Jenner's House: www.jennermuseum.com

The Voltaic Pile

Fara, Patricia. *An Entertainment for Angels: Electricity in the Enlightenment.* Cambridge, UK: Icon, 2003.

Pancaldi, Giuliano. *Volta: Science and Culture in the Age of Enlightenment,* Princeton, N.J.; Woodstock, NY: Princeton University Press, 2003.

No Laughing Matter

Holmes, Richard. *The Age of Wonder.* London, UK: HarperCollins, 2008

Jay, Mike. *The Atmosphere of Heaven: The Unnatural Experiments of Dr. Beddoes and His Sons of Genius.* New Haven, CT; London, UK: Yale University Press, 2009.

Zuck, David, Peter Ellis and Alan Dronsfield. "Nitrous Oxide: Are You Having a Laugh?" *Education in Chemistry,* March 2012.

Infrared Light

Clegg, Brian. *Light Years: An Exploration of Mankind's Enduring Fascination with Light.* London, UK: Piatkus, 2001.

Herschel Space Observatory: http://herschel.cf.ac.uk

The Air Up There

Conner, Susan, and Linda Kitchen. *Science's Most Wanted: The Top 10 Book of Outrageous Innovators, Deadly Disasters, and Shocking Discoveries.* Washington, DC: Potomac Books, Inc., 2002.

Holmes, Richard. *Falling Upwards: How We Took to the Air: An Unconventional History of Ballooning.* London, UK: William Collins, 2013.

The Great Falls of the Missouri

Ambrose, Stephen E. *Undaunted Courage: The Pioneering First Mission to Explore America's Wild Frontier.* London, UK: Pocket Books, 2003.

Lewis and Clark's Historic Trail: http://lewisclark.net

Electromagnetism

Berkson, William. *Fields of Force: The Development of a World View from Faraday to Einstein.* Oxford, UK: Routledge, 2014.

Cunningham, Andrew, and Nicholas Jardine, eds. *Romanticism and the Sciences.* Cambridge, UK: Cambridge University Press, 1990.

Magnet Lab, National High Magnetic Field Laboratory: www.magnet.fsu.edu

The First Dinosaur Fossils

Dean, Dennis R. *Gideon Mantell and the Discovery of Dinosaurs*. Cambridge, UK: Cambridge University Press, 1999.

Moody, Richard, ed. *Dinosaurs and Other Extinct Saurians: A Historical Perspective*. London, UK: Geological Society of London, 2010.

The Unpublished Journal of Gideon Mantell: www.brighton-hove-rpml.org.uk/HistoryAndCollections/aboutcollections/naturalsciences/Documents/mantell_journal.pdf

Natural Selection

About Darwin: www.aboutdarwin.com

Darwin Online: darwin-online.org.uk

The Darwin Project: www.darwinproject.ac.uk

The Agent of Infection

Brody, Howard, Russell Rip, Michael, Vinten-Johansen, Peter, Paneth, Nigel, and Rachman, Stephen. "Map-making and Myth-making in Broad Street: The London cholera epidemic, 1854." *The Lancet*, Vol. 356. July, 2000.

Koch, Tom. "The Map as Intent: Variations on the Theme of John Snow." *Cartographica*, Volume 39, No.4, Winter 2004.

The Source of the Nile

Jeal, Tim. *Explorers of the Nile: The Triumph and Tragedy of a Great Victorian Adventure*. London, UK: Faber, 2012.

Moorehead, Alan. *The White Nile*. London, UK: Hamilton, 1971.

Plumb, J. H. "The Search for the Nile." *History Today*, Vol. 2 (11), 1952.

The Laws of Inheritance

Marantz Henig, Robin. *A Monk and Two Peas: The Story of Gregor Mendel and the Discovery of Genetics*. London, UK: Phoenix, 2001.

Mawer, Simon. *Gregor Mendel: Planting the Seeds of Genetics*. New York, NY: Abrams, in association with the Field Museum, Chicago, 2006.

Reeve, Eric C. and Black, Isobel. *Encyclopedia of Genetics*. Oxford, UK: Taylor & Francis, 2001.

Radio Waves

Fowler, Michael. "The Photoelectric Effect." Modern Physics, University of Virginia: http://galileo.phys.virginia.edu/classes/252/photoelectric_effect.html

James Clerk Maxwell Foundation: www.clerkmaxwellfoundation.org

The Polyphase Motor

Cheney, Margaret. *Tesla: Man Out of Time*. Englewood Cliffs, London, UK: Prentice-Hall, 1981.

Seifer, Marc J. *Wizard: The Life and Times of Nikola Tesla*. Secaucus, NJ: Carol Publishing Group, c.1996.

Tesla, Nikola. *My Inventions: The Autobiography of Nikola Tesla*. Williston, VT: Hart Brothers, 1982.

X-rays

Michette, A. *X-Rays: The First Hundred Years*. West Sussex, UK: John Wiley & Sons, 1996:

Segrè, Emilio. *From X-rays to Quarks*. Berkeley, CA: University of California, 1980.

Radioactivity

Goldsmith, Barbara. *Obsessive Genius: The Inner World of Marie Curie*. London, UK: Phoenix, 2005.

Malley, Marjorie Caroline. *Radioactivity: A History of a Mysterious Science*. Oxford, UK: Oxford University Press, 2011.

The Electron

Pais, Abraham. *Inward Bound: Of Matter and Forces in the Physical World*. Oxford, UK: Clarendon, 1986.

Classic Chemistry: https://web.lemoyne.edu/~giunta/EA/THOMSONann.html

Making Sense of Matter

Reeves, Richard. *A Force of Nature: The Frontier Genius of Ernest Rutherford*. New York, NY: W. W. Norton & Co., 2008.

Gravitational Lensing and Relativity

Feynman, Richard P. *Six Not-so-easy Pieces: Einstein's Relativity, Symmetry, and Space-time*. New York, NY: Addison-Wesley Pub, 1997.

Kennefick, Daniel. "Testing Relativity from the 1919 Eclipse—A Question of Bias." *Physics Today*, March 2009.

Pais, Abraham. *Subtle is the Lord: The science and Life of Albert Einstein*. Oxford, UK: Oxford University Press, 1982.

Penicillin

Brown, Kevin. *Penicillin Man: Alexander Fleming and the Antibiotic Revolution*. Stroud, UK: Sutton, 2004.

Chemical Heritage Foundation: www.chemheritage.org

Nuclear Fission

Clark, Ronald W. *The Greatest Power on Earth: The Story of Nuclear Fission*. London, UK: Sidgwick & Jackson, 1980.

Rhodes, Richard. *The Making of the Atomic Bomb*. London, UK: Penguin, 1988.

The Lascaux Caves

Morriss-Kay, Gillian M. "The Evolution of Human Artistic Creativity." *Journal of Anatomy*, Vol. 216 (2), 158–176, February 2010.

Lascaux official website: www.lascaux.culture.fr

The Double Helix

Aldridge, Susan. "The DNA Story." *Chemistry World*, April 2003.

Hunter, Graeme K.. *Vital Forces: The Discovery of the Molecular Basis of Life*. New York, NY: Elsevier, 2000.

Watson, James D. *The Double Helix*. London, UK: Penguin, 1971.

The Butterfly Effect

Gleick, James. *Chaos: Making a New Science*. New York, NY: Viking, 1987.

Gribbin, John. *Deep Simplicity: Chaos, Complexity and the Emergence of Life*. London, UK: Allen Lane, 2004.

Spinning Neutron Stars

Bell Burnell, S. Jocelyn. "Little Green Men, White Dwarfs or Pulsars?" *Annals of the New York Academy of Science*, vol. 302, 685–689, December 1977.

National Radio Astronomy Observatory: www.nrao.edu

"Lucy," the Hominid Skeleton

Johanson, Donald, and Maitland Edey. *Lucy: The Beginning of Humankind*. London, UK: Simon & Schuster, 1981.

O'Neil, Dennis. Early Human Evolution: http://anthro.palomar.edu/homo/

Human Origins, Smithsonian National Museum of Natural History: http://humanorigins.si.edu

Life in the Depths

Ballard, Bob. *The Eternal Darkness*. Princeton, NJ: Princeton University Press, 2000.

Kaharl, Victoria. *Water Baby: The Story of Alvin*. Oxford, UK: Oxford University Press, 1991.

The Discovery of Hydrothermal Vents 25th Anniversary:

www.divediscover.whoi.edu/ventcd/vent_discovery

The First Exoplanet

Fischer, Debra. "Prowling for Planets." *Mercury Magazine*, vol. 29 (4), July/August 2000.

Jayawardhana, Ray. *Strange New Worlds: The Search for Alien Planets and Life Beyond Our Solar System*. Princeton, NJ: Princeton University Press, 2013.

Planet Quest, NASA Jet Propulsion Laboratory: http://planetquest.jpl.nasa.gov

The Higgs Boson

Lederman, Leon. *The God Particle: If the Universe Is the Answer, What Is the Question?* New York, NY: Dell Publishing, 1993.

Sample, Ian. *Massive: The Higgs Boson and the Greatest Hunt in Science*. London, UK: Virgin, 2013.

CERN: http://home.web.cern.ch